Privacy and Safety in Online Learning

Privacy

and

Safety

in Online Learning

Edited by
Denise FitzGerald Quintel and Amy York

Middle Tennessee State University

Privacy and Safety in Online Learning

© 2023 the Authors
Published by MT Open Press at Middle Tennessee State University · Murfreesboro

Each chapter has been peer-reviewed, and the entire book has been externally reviewed as part of the quality review process. While the publisher and authors have used good faith efforts to ensure the quality of information in this work are accurate, the publisher and authors disclaim all responsibility for errors or omissions. Use of the information in this work is at your own risk. All URL links worked at the time of publication.

 This work is licensed under a Creative Commons Attribution-NonCommercial-NoDerivatives 4.0 International License.

Identifiers
ISBN (paperback) 979-8-9871721-1-7
ISBN (digital PDF) 979-8-9871721-0-0
DOI: 10.56638/mtopb00123

Library of Congress Data
LC Control Number: 2022950399
Library of Congress Subject Headings: Distance education — privacy — data privacy

Press Operations
MT Open Press, an imprint of Digital Scholarship Initiatives at the James E. Walker Library, Middle Tennessee State University. https://openpress.mtsu.edu

Copyeditor: Emma Sullivan, Cover design: A.Miller with cover image adaptions from @littleliondogs and Canva. Typeset in Garamond, Rockwell, Sabon and Bembo.

Digital version (PDF) available at https://openpress.mtsu.edu

Print-on-demand version (paperback) available at https://www.lulu.com/spotlight/mtop

Suggested Citation
Quintel, D. F. & York, A. (2023). Privacy and safety in online learning. MT Open Press, Middle Tennessee State University. https://doi.org/10.56638/mtopb00123

 Middle Tennessee State University does not discriminate against students, employees, or applicants for admission or employment on the basis of race, color, religion, creed, national origin, sex, sexual orientation, gender identity/expression, disability, age, status as a protected veteran, genetic information, or any other legally protected class with respect to all employment, programs, and activities sponsored by MTSU. More information is available at www.mtsu.edu/iec

Table of Contents

Preface — vii

Introduction

Why Privacy, Why Now? — 3

Denise FitzGerald Quintel and Amy York

Part 1: What's the State of Privacy in Online Education?

The Importance of Data Privacy and Security During Emergency Remote Learning — 13

Emma Antobam-Ntekudzi

"At the Cost of My Well-being": Exploring Trans, Non-binary, and Gender-Diverse Students' Experiences of Online Learning — 29

Maddie Brockbank, Wil Fujarczuk, Christian Barborini, and Yimeng Wang

What Privacy? Online Privacy Culture and the Role of Libraries in Digital Information Literacy — 49

Hannah Lee

Part 2: How Do We Take Care of Our Students, Ourselves?

Professional Identity and Digital Diligence — 67

Angela Dixon and Amy Stalker

Online Harassment in Elementary Schools — 81

Rebecca Taylor

Ask What You Want; We Don't Know Who You Are: Live Chat, Library Anxiety, and Privacy in an Academic Library — 95

Bridgette Sanders, Jon B. Moore, and Kimberly Looby

Part 3: How Can We Transform Our Pedagogy with Privacy in Mind?

Visible Bruises: Domestic Violence and Trauma-Informed Instruction in Remote Learning Environments — 121

Jennifer Lynn Reichart

Pedagogy of Privacy: Inclusive Teaching and Disclosures of Disability — 137

Sarah Whitwell and Samantha Clarke

Remote Learning Environments for Students who are Academically at-risk, Non-traditional, or from Diverse Backgrounds — 151

Christina M. Cobb and Meredith Anne (MA) Higgs

Part 4: What Tools or Resources Can We Use?

"Have You Seen My Cartoon Yet?": Objectives on Managing Student Projects in an Online STEAM Program — 165

Kristen Vogt Veggeberg

The Cost of Respect? Surprisingly Little — 173

Joseph Kennedy and Albert Kagan

Privacy in the Online Writing Center — 191

James Hamby

Artificial Intelligence for Privacy Conservation in Remote Learning — 201

Hongbo Zhang, Lei Miao, Jia-Xing Zhong, and Aimin Yan

About the Authors — 221

Acknowledgements — 229

Index — 230

Preface

***

While the Covid-19 pandemic created a mainstream conversation about privacy and safety issues in online learning, it is important to acknowledge these issues will remain even when the pandemic ends.

***

This collection features essays, case studies, and pedagogical approaches that explore how educators managed the privacy, security, and safety concerns that rushed into our lives as we shifted into emergency remote learning in 2020. While the COVID-19 pandemic brought this concern into focus, privacy issues with online learning continue to exist alongside us and our students.

This book provides readers insight into the current state of privacy issues, describes the challenges and rewards of developing more privacy-focused learning environments, and presents several resources and tools that readers can bring to their own teaching practices.

Representing a variety of perspectives from K-12, higher education, and libraries, contributors describe the challenges they encountered and offer solutions to help ensure the safekeeping of students' online lives. How do we navigate these online environments, who collects our data, and how can we protect our most vulnerable populations?

Keywords: privacy, online learning, educational technology, digital pedagogy, emergency remote learning, COVID-19

Introduction

Why Privacy, Why Now?

Denise FitzGerald Quintel and Amy York

In many ways, we are all still reeling from 2020. Caregivers, parents, children, educators, and virtually everyone took on numerous roles and responsibilities while coping with losses, grief, and trauma that we all seemed to share collectively. There was a consistent need to connect and find community when the world seemed to shut down around us, with so many navigating spaces that seemed impossible to pass.

Unsurprisingly, the start of the COVID-19 pandemic altered the educational landscape on a tremendous scale. The rush to emergency remote learning was one of the most significant challenges that educators, parents, and students faced, and it is something that still impacts us. Even though schools have returned to in-person classes, online platforms hastily adopted in 2020 remain used as course management, communication, or content delivery tools. Privacy issues related to education are not new, but the sudden shift to online learning brought these concerns into sharp focus for many parents, educators, administrators, and researchers.

The objective of this book is to reflect on the unintended breaches of privacy, safety, and security that occurred during 2020 and how those events continue to shape online educational spaces. Within these chapters, contributors examine their own teaching experiences and propose solutions for more responsible use of online platforms and tools. This book documents how educational institutions approach privacy. It describes initiatives implemented in response to online learning and contributes to the growing discussion of how privacy and surveillance impact our users, especially students from our most vulnerable populations.

In 2020, Pew Research presented survey results illustrating a growing problem with how private companies collect our data. In the prior year, three-quarters of Americans (72%) believed that private companies utilized almost all their data, and nearly half (47%) believed the government was also surveilling their data (Auxier et al., 2019). The COVID-19 pandemic response created more concerns as government and private companies used personal devices for tracking individuals testing positive for the virus. However, while most

Americans feel uneasy about their data being collected, many also think that privacy protection concepts are too complex to understand or implement on our own (Auxier, 2019). As editors, we could not anticipate the sheer number of privacy issues that would continue to impact news stories while compiling this book between 2021 and 2022.

Although there have been reports on the data collection business for a while (Valentin-DeVries et al., 2018), we have been on high alert for location tracking and collecting personal health data this past year. Most notably, after the Supreme Court overturned Roe V. Wade, many people deleted menstrual cycle-tracking apps from their phones (Kwong et al., 2022). Still, those actions sparked more discussion into how many ways our data is collected and identifiable whenever we use a smartphone, website, or Google search (Hill, 2022). Reporters investigated data brokers and demonstrated the incredible ease and surprisingly low cost involved with purchasing and then deanonymizing aggregate data (Cox, 2022).

There were legal wins against proctoring software in the education world, declaring room scans unconstitutional (Bowman, 2022). More reports surfaced on how online proctoring poses significant problems for users with disabilities (Brown, 2022). Additionally, student activity monitoring software, which became more widely used during the shift to online learning, is still being used by institutions at a global level despite numerous red flags (CDT, 2021; Singer, 2022). Even the youngest children and their families are at risk, as we are now seeing reports on how daycare apps are rife with security and data issues (Gruber et al., 2022; Hancock, 2022).

We saw library service providers purchased by data analytics companies for billions (USD), moving away from traditional publishing companies and finding new ways to monetize data points in addition to scholarly research (Lamdan et al., 2021). In some cases, we saw library vendors engage in partnerships with government agencies, granting access to the personal data of millions (Coldeway, 2022; Lamdan, 2019; Lui, 2022). We saw numerous data breaches in school districts and higher education institutions (Bamforth, 2022; Freed, 2022; Johnson, 2022). Even if these schools paid the ransom, there would be no guarantee that these institutions would recover their data (Klein, 2022; Mahendru, 2022; Page, 2022; Singer, 2022). Privacy is far from a new or novel concept in education, but as more of these stories bubbled to the surface, it was clear that concerns were valid and piqued public interest.

While many families balked at the surveillance features in school-provided technology (Ceres, 2022), some chose to double down on surveillance measures for their children in 2020. We witnessed multiple states enact legislation to censor library materials (Iowa 2176; Oklahoma SB1142; Indiana SB17; Idaho HB666; Tennessee HB/SB1944). At a local level, even school boards censored materials (Mangrum, 2021). In response to legislation, we watched a company, through their technology, attempt to give parents unfettered access to their child's library history without the consent of their children. Included in that technology would also be an effortless way for parents to restrict materials, such as anything tagged for LGBTQ content (Ellis, 2022). Some may argue that libraries should use surveillance measures for protection. Nevertheless, with surveillance technologies, it is essential to look closely at how companies and their parent companies engage in business (Gallagher, 2020; Krapiva & Micek, 2020).

As library professionals, we want to point out that privacy is a core value of our profession. Any attempt to censor, restrict, monitor, or suppress the free flow of ideas is antithetical to intellectual freedom. Groups who push for censorship in libraries to "protect" children directly oppose what safe spaces for learning, creating, and self-expression look like for all children.

> Privacy is essential to the exercise of free speech, free thought, and free association. Lack of privacy and confidentiality chills users' choices, thereby suppressing access to ideas. The possibility of surveillance, whether direct or through access to records of speech, research, and exploration, undermines a democratic society.
> - Privacy: An Interpretation of the Library Bill of Rights (ALA, 2019)

***

The chapters collected in this book describe a wide array of privacy issues in online and remote learning environments. Our contributors go beyond the practical takeaways for keeping information and data safe. We see how educators, librarians, and administrators share an underlying motivation to protect their students while safeguarding students' autonomy. The authors capture the frantic energy many educators experienced as we shifted to emergency remote learning and how it shaped and continues to influence these online spaces. Their experiences are as varied as their online spaces, as we hear from writing centers, out-of-school elementary programs, libraries, middle schools, and universities.

We have organized the book into different sections, each attempting to answer an overarching question. Section I provides an overview of what institutions currently do to address privacy concerns. Contributors share how to build collaborative, safe learning policies and reveal the shortcomings of the Family Educational Rights and Privacy Act (FERPA). One author shares her approach to collaborative policy building, where all voices and viewpoints will have a seat at the table to craft ways to address privacy issues. Another addresses online privacy culture through the lens of an academic librarian, which involves increasing accountability at a system level rather than at the individual.

Section II looks closely at how we can protect our students and ourselves as educators. Authors bring the voices of transgender, non-binary, and gender-diverse students to the discussion and ask readers to listen to how we could improve their online experiences. Authors provide ways to help ensure instructor safety as lines between personal and professional life often blur during remote instruction. The chapters cover doxing, online bullying, and library-induced anxiety.

Section III examines how others have built or transformed their online pedagogy to incorporate privacy and safety concerns. We ask readers to consider what privacy looks like for marginalized groups and at-risk students and how we can improve the care for our students and ourselves. We hear about building authentic connections with our students while protecting their private lives. We learn how a trauma-informed pedagogy can help students and how privacy included in universal design learning benefits everyone.

Lastly, in section IV, the authors provide several tools and resources that one can implement into their online instructional spaces. The authors discuss privacy tools ranging from artificial intelligence (AI) methods to proctoring alternatives and best practices for storing data. Many authors share valuable privacy-focused resources and tools that an educator can consider, ranging from beginner to expert-level experience for implementation.

***

While the idea for this book came from the experiences that we, as co-editors, had as parents, librarians, caregivers, and on-call teachers during the early stages of the COVID-19 pandemic, our original book title solely referenced remote learning. As readers will see in the following chapters, even though initial experiences occurred during emergency remote teaching (ERT), the lessons and resources shared go beyond any single environment and can inform several types of instruction and educational backgrounds. Like the legal realm, privacy in educational settings is a concept that cannot be defined as one thing but contains many ideas and definitions (Hertzog, 2021).

When we sent out the initial call for chapters, we were unsure how receptive scholars, practitioners, and educators would be.

The response was stunning.

While this book will not have all the answers to your questions, it provides a great starting point for those interested in addressing privacy, safety, and security concerns in their own online and remote educational environments.

"All of this to say that:
You deserve safety & agency
You deserve more & better
Then what is offered."
- Library Freedom Project FINSTA Project, 2020

References

American Library Association (ALA). (2019, June 24). Privacy: An interpretation of the Library Bill of Rights. https://www.ala.org/advocacy/intfreedom/librarybill/interpretations/privacy

American Library Association. (2021, October 20). Core values. Advocacy, Legislation & Issues. https://www.ala.org/advocacy/privacy/values

Auxier, B. (2020, August 27). How Americans see digital privacy issues amid the COVID-19 outbreak. Pew Research Center. https://www.pewresearch.org/fact-tank/2020/05/04/how-americans-see-digital-privacy-issues-amid-the-covid-19-outbreak/

Auxier, B., Rainie, L., Anderson, M., Perrin, A., Kumar, M., & Turner, E. (2019, November 15). Americans and privacy: Concerned, confused and feeling lack of control over their personal information. Pew Research Center. https://www.pewresearch.org/internet/2019/11/15/americans-and-privacy-concerned-confused-and-feeling-lack-of-control-over-their-personal-information/

Bamforth, E. (2022, May 2). After ransomware, Austin Peay moves ahead with finals. EdScoop. https://edscoop.com/austin-peay-state-university-ransomware-finals-petition/

Bowman, E. (2022, August 26). Scanning students' rooms during remote tests is unconstitutional, judge rules. NPR. https://www.npr.org/2022/08/25/1119337956/test-proctoring-room-scans-unconstitutional-cleveland-state-university

Brown, L. X. Z. (2021, June 23). How automated test proctoring software discriminates against disabled students. Center for Democracy and Technology. https://cdt.org/insights/how-automated-test-proctoring-software-discriminates-against-disabled-students/

Ceres, P. (2022, October 10). How to protect yourself if your school uses surveillance tech. Wired. https://www.wired.com/story/how-to-protect-yourself-school-surveillance-tech-privacy/

Cox, J. (2022, May 3). Data broker is selling location data of people who visit abortion clinics. VICE. https://www.vice.com/en/article/m7vzjb/location-data-abortion-clinics-safegraph-planned-parenthood

Coldewey, D. (2022, June 9). Records show ICE uses LexisNexis to check millions, far more than previously thought. TechCrunch. https://techcrunch.com/2022/06/09/records-show-ice-uses-lexisnexis-to-check-millions-far-more-than-previously-thought/

Ellis, D. (2022, April 4). Technology for parent monitoring of Student Library use is being developed by Follett: This Week's book Censorship News April 1, 2022. Book Riot. https://bookriot.com/book-censorship-news-april-1-2022/

Freed, B. (2022, January 14). Albuquerque, New Mexico, schools closed after cyberattack. State Scoop. https://statescoop.com/albuquerque-new-mexico-schools-closed-after-cyberattack/

Gallagher, R. (2020, August 28). Belarusian officials shut down internet with technology made by U.S. firm. Bloomberg.com. https://www.bloomberg.com/news/articles/2020-08-28/belarusian-officials-shut-down-internet-with-technology-made-by-u-s-firm?leadSource=uverify+wall

Gallagher, R. (2021, January 26). Private equity firm Francisco Partners profits from surveillance, censorship. Bloomberg.com. https://www.bloomberg.com/news/features/2021-01-26/private-equity-firm-francisco-partners-profits-from-surveillance-censorship

Grant-Chapman, H., Laird, E., & Venzke, C. (2022, May 19). Student activity monitoring software: Research insights and recommendations. Center for Democracy and Technology. https://cdt.org/insights/student-activity-monitoring-software-research-insights-and-recommendations/

Gruber, M., Höfig, C., Golla, M., Urban, T., & Große-Kampmann, M. (2022). "We may share the number of diaper changes": A privacy and security analysis of Mobile Child Care Applications. Proceedings on Privacy Enhancing Technologies, 2022(3), 394–414. https://doi.org/10.56553/popets-2022-0078

Hancock, A. (2022, June 27). Daycare apps are dangerously insecure. Electronic Frontier Foundation. https://www.eff.org/deeplinks/2022/06/daycare-apps-are-dangerously-insecure

Hartzog, W. (2021). What is privacy? That's the wrong question. The University of Chicago Law Review, 88(7), 1677–1688. https://www.jstor.org/stable/27073959

Hill, K. (2022, July 11). Deleting your period tracker won't protect you. New York Times. https://www.nytimes.com/2022/06/30/technology/period-tracker-privacy-abortion.html

House Bill 666. (2022, July 1). 66th Legislature of the State of Idaho. https://legislature.idaho.gov/sessioninfo/2022/legislation/H0666/

House Bill 1944 (2022, March 30). 112th General Assembly of the State of Tennessee. https://www.capitol.tn.gov/Bills/112/Bill/HB1944.pdf

House File 2176. (2022, February 1). 89th General Assembly of the State of Iowa. https://www.legis.iowa.gov/legislation/BillBook?ga=89&ba=HF2176

Klein, A. (2022, March 31). What schools can learn from the biggest cyberattack ever on a single district. Education Week. https://www.edweek.org/technology/what-schools-can-learn-from-the-biggest-cyberattack-ever-on-a-single-district/2022/03

Krapiva, N., & Micek, P. (2020, September 4). Francisco Partners-owned Sandvine profits from shutdowns and oppression in Belarus. Access Now. https://www.accessnow.org/francisco-partners-owned-sandvine-profits-from-shutdowns-and-oppression-in-belarus/

Kwong, E., Ramirez, R., & Cirino, M. (2022, January 18). When tracking your period lets companies track you. NPR. https://www.npr.org/2021/12/29/1068930998/when-tracking-your-period-lets-companies-track-you

Lamdan, S. (2019, November 13). Librarianship at the crossroads of ICE Surveillance. In the Library with the Lead Pipe. https://www.inthelibrarywiththeleadpipe.org/2019/ice-surveillance/

Lamdan, S., Montoya, R., Swauger, S., & Halperin, J. R. (2021, December). The conquest of ProQuest and Knowledge Unlatched: How recent mergers are bad for research and the public. Invest in Open Infrastructure. https://investinopen.org/blog/the-conquest-of-proquest-and-knowledge-unlatched-how-recent-mergers-are-bad-for-research-and-the-public/

Library Freedom Project. (2021, April). The FINSTA project: What educational tech knows. Library Freedom Project. https://libraryfreedom.org/finsta-project/

Lui, Y. (2022). LexisNexis and I.C.E.: An examination of LexisNexis's human rights responsibilities. Journal of International Law and Politics, 54(23), 69–84. https://www.nyujilp.org/lexisnexis-and-i-c-e-an-examination-of-lexisnexiss-human-rights-responsibilities/

Mahendru, P. (2022, July 12). The state of Ransomware in education 2022. Sophos News. https://news.sophos.com/en-us/2022/07/12/the-state-of-ransomware-in-education-2022/

Mangrum, M. (2021, October 21). Tennessee librarians speak out against Chattanooga school board member's attempt to ban books. The Tennessean. https://www.tennessean.com/story/news/education/2021/10/21/tennessee-librarians-speak-out-against-chattanooga-school-board-members-attempt-have-books-banned-sc/6119874001/

Page, C. (2021, November 22). US education software company exposed personal data of 1.2M students. TechCrunch. https://techcrunch.com/2021/11/22/smarterselect-exposed-millions-student-data/

Senate Bill 17. (2022, January 28). 122nd General Assembly of the State of Indiana. https://iga.in.gov/documents/221c1669

Senate Bill 1142. (2021, December 16). 58th Legislature of the State of Oklahoma. http://www.oklegislature.gov/BillInfo.aspx?Bill=SB1142&Session=2200

Senate Bill 1944. (2022, April 6). General Assembly of the State of Tennessee. https://www.capitol.tn.gov/Bills/112/Bill/SB1944.pdf

Singer, N. (2022, July 31). A cyberattack illuminates the shaky state of student privacy. New York Times. https://www.nytimes.com/2022/07/31/business/student-privacy-illuminate-hack.html

Valentin-DeVries, J., Singer, N., Keller, M. H., & Krolik, A. (2018, December 10). Your apps know where you've been and can't keep a secret. New York Times. https://ezproxy.mtsu.edu/login?url=https://www.proquest.com/newspapers/your-apps-know-where-youve-been-cant-keep-secret/docview/2153542863/se-2

/ *Part 1: What's the State of Privacy in Online Education?*

The Importance of Data Privacy and Security During Emergency Remote Learning

Emma Antobam-Ntekudzi

The COVID-19 pandemic forever changed the world. The virus' rapid spread forced federal and local governments to enact quarantine mandates. On March 11, 2020, the Center for Disease Control and Prevention (CDC) (2022) announced COVID-19 as a pandemic. Two days later the United States declared an official nationwide emergency. Institutions were required to shut down and persons deemed non-essential participated in quarantine. Remote working became the standard, thus affecting all aspects of individual lives and institutions, especially education. Primarily in-person universities and colleges across the world scrambled to address the COVID-19 health concerns, comply with local shutdown rules, and attempt to continue providing an education to millions of students. Having no other option, faculty and other instructors were apprehensively thrust into the world of solely online teaching and learning (Paris et al., 2021). Instructors became resourceful in their techniques to quickly provide content online for their students. Unfortunately, this massive shift left little room to assess student information privacy and security concerns in a new non-traditional online environment. Easy-to-use and free teaching tools were adopted without these considerations. After two years, institutions can reflect on compromises made to data privacy and security in response to the COVID-19 crisis, particularly since a fully online presence opened some institutions to hacking vulnerabilities. Moving forward, students, instructors, administrators, and information technology staff should have a seat at the table when outlining privacy and security policies, during educational technology tool selection, and ensuring safe online learning.

Traditional Online Learning vs. Emergency Remote Learning

Online learning is not a new phenomenon. Prior to the pandemic, online learning existed as part of various distance learning programs (Shearer et al., 2020). For years, colleges and universities that are established as online institutions offered a modality of learning to students who preferred a remote education. It can be referred to by other names such as blended learning, mobile learning, and distributed learning (Nørgård, 2021). This form of teaching requires all aspects of a course to be available digitally. Depending on the parameters of each course, a class can be held asynchronously or synchronously. Students depend on access to technology devices to attend weekly classes, participate, and complete coursework. In this modality, face-to-face in-person engagement with classmates and/or the professor is usually non-existent. However, in-person components can be at the discretion of the instructor. Professors may meet with students during office hours in person and/or require group projects where students possibly gather in person to complete assignments. Online learning programs may include courses that follow a hybrid model, encompassing the blending of online and offline, formal, and informal (Nørgård, 2021) teaching and learning. Such established online learning programs support lifelong learning because of their flexibility for students who encounter a multitude of social and personal challenges that reduce the ability to succeed in a traditional face-to-face learning environment (Nørgård, 2021). Distance learning constituted a niche learning modality (Shearer et al., 2020) prior to the pandemic. COVID-19 brought this style of teaching and learning to the forefront where it was adopted rapidly (Kim, 2020) and under duress.

The term online learning is sometimes used to describe the kind of teaching and learning enacted in reaction to the COVID-19 pandemic. However, the rushed response, unpreparedness of instructors, and desperation of higher education institutions to resume their courses culminated in emergency remote learning (Karakaya, 2020). Traditional online learning includes established theory, pedagogy, assessment, conceptual frameworks, best practices, and defined terms (Nørgård, 2021). In addition to this, there is abundant literature on teaching and learning online (Ibacache et al., 2021). Emergency remote learning is different because of the nature in which it arises (Karakaya, 2020). It is a temporary solution that provides an opportunity for continuity of education over a certain amount of time. Traditional online learning is not designed for provisional purposes. Understandably, the accelerated push to move everything online was not accompanied by proper support (Walsh et al., 2021). A lack of formal online teaching training for instructors accustomed to teaching predominantly in-person contributed to poor implementation of courses, particularly in the early stages of the pandemic. Limited training in universal design further weakened the quality of these online courses (Nørgård, 2021).

Teaching online was unfamiliar for a major portion of instructors (Ibacache et al., 2021) and face-to-face institutions. Some educational technology was already implemented in face-to-face classes pre-pandemic and used widely. An example of this is the learning management system (LMS) Blackboard. Unless a course was fully online prior to the pandemic, Blackboard was often used as supplemental to in-person teaching. It functioned as a central location for housing required materials, allowing for assignment submission, sharing announcements from the professor, and providing a location for communication outside of the in-person classroom. A major feature and benefit of traditional face-to-face courses remained the in-person interaction and courses were developed around the understanding of this modality. Upon COVID-19's arrival, the lack of time to train instructors did not allow for formal pedagogical teaching in online course development. Many were forced to figure out how to convert all face-to-face courses to 100% online within 24 hours. Assessments designed for in-person courses needed modification; instructors were still required to fulfill the learning outcomes of an academic department in this unfamiliar modality.

Technical Side of Security

The use of online tools to aid in the dissemination of emergency remote learning highlights the resilience and dedication of instructors during a health disaster. However, a discussion of privacy and security is necessary when educational technology or tools, not designed for education, are adopted. Instructors use educational tools adopted by the institution with the assumption they have been vetted in some form. The understanding is that a license between the vendor and institution exists. When individual instructors decide to seek and use tools not supplied by the institution, especially during an emergency, there is less chance that security and privacy outlines are investigated. For instance, cloud computing is important for online learning (Ali, 2021) and is embedded into some educational technology. Tools such as Dropbox, Google Drive, and OneDrive are popular cloud storage options used by colleges and universities. Users' personal and professional information is maintained in these spaces. This can pose a risk to security and privacy if licenses or terms of service with vendors do not provide clarity on data ownership, or they may include language that releases vendors from liabilities (Paris et al., 2021). Instructors who adopt Google Drive may not investigate the service's privacy/security policies. It is important to consider to whom the information belongs (the individual, the third-party vendor, or the institution), with whom it is shared, and what happens to the data once the contract is terminated (Gelpi, 2020). Institutions and instructors should be aware of whether the cloud vendor provides support (Dennen & Burner, 2017) in instances where data is breached.

Student Privacy

As instructors worked through emergency remote learning, the ability to create an online community in the classroom became a challenge. Some turned to well-known mobile applications such as WhatsApp to aid classroom communication, further collaborative work, and maintain engagement with the course (Tarisayi & Munyaradzi, 2021). Though such applications are familiar to students and perhaps used personally, questions should be raised regarding their security in a formal learning setting. Applications usually contain click-wrap agreements (Gelpi, 2020) that request access to a user's contacts. It will also allow the individual the opportunity to deny or accept the application for download. Policies outlined within click-wrap terms are vague regarding the use of data or no information is provided on whether the data is destroyed (Paris et al., 2021). In addition to this, if an instructor has required students to use the application, there is no consent on the part of the student. Even with the existence of a digital agreement and request, it is in the best interest of the student to agree otherwise their success in the course could be jeopardized.

Student concerns about privacy and security may limit participation and negatively impact trust (Kim, 2020). There is little to no guaranteed protection when using free online tools. Applications and free tools are susceptible to hacking when used on unsecured devices (Kim, 2020). Information technology departments at colleges may implement firewall protections, but this primarily works when a student or instructor is on campus using the college's Wi-Fi. COVID-19 forced instructors and students to remain off-site while engaging in teaching and learning. This required the use of internet access that is likely not protected, such as data for personal devices and public or home Wi-Fi. There are varying studies that reveal differences in student perceptions of their data security. One study found that students were not concerned if professors and institutions utilized their data for research and education purposes (Vu et al., 2019). Institutions and instructors can guarantee students that they mean to use their data for approved purposes, but they cannot guarantee this is being done by third-party vendors. Regardless of student perceptions on this topic, being transparent and trying to improve privacy can impact student satisfaction (Williams et al., 2019). This applies not only to the classroom but also to college services.

The use of video conferencing tools raises concerns about not only privacy protections but also surveillance issues. Traditionally, those from marginalized groups have experienced forms of surveillance in their communities and are not usually afforded the same levels of privacy as those of dominant identities (Paris et al., 2021). Institutions that serve Black Indigenous People of Color (BIPOC) populations must be aware of how privacy intrusion can harm those from marginalized communities. Perceived surveillance can re-traumatize vulnerable students. Therefore, visitors or guests to an online class should be announced to students prior to their arrival. Instructors can impair trust when visitors are allowed into an online session and left unidentified. Instructors using the Zoom Video Communications conferencing platform must ensure protocols are in place for situations known as Zoom-bombing, where uninvited users take advantage of weak security protocols (Elmer et al., 2021). In 2020, a lawsuit filed against the Zoom Video Communications company alleged that it "sold user data to Facebook" (Brooks, 2020). The company's CEO used this opportunity to address security lapses that created an environment for Zoom-bombing to occur easily (Brooks, 2020). Since the latter half of 2020, instances of Zoom-bombing reduced due to the reaction from the company. Although, it remains important for instructors using this platform to be vigilant and make concessions with students who exercise privacy rights. Students who wish to maintain a camera-off participation style during synchronous online learning should not be penalized with a negative participation grade. Students who choose to reveal their faces on camera and full names should be provided with assurances that this identifiable information will not be exploited (Kim, 2020). Compromised student privacy can be detrimental to student learning (Vu et al., 2019).

FERPA

Institutions have the option to officially designate a widely used and necessary third-party tool, like Zoom or Blackboard. This allows the vendor access to sensitive information in a limited capacity (Gelpi, 2020). Even with the school's official designation, vendor agreements must still indicate which party receives direct control and maintenance of education records, images, and recordings. The responsibility is on the institution or instructor to ensure that an adopted educational tool is following the Family Educational Rights and Privacy Act (FERPA). FERPA, enacted in 1974 (CDC, 2018), was designed to give parents and eligible students over 18 control of their education records. It prohibits the release or disclosure of personally identifiable information without written consent, although there are some entities that can legally obtain student educational records without consent and under specific circumstances (U.S. Department of Education, 2021). These include other schools to which a student will transfer, police authorities, disclosures in response to a subpoena, and accrediting organizations (U.S. Department of Education, 2021). FERPA provides basic guidelines for protection and is part of the conversation around ethics in data analytics in education. FERPA cannot cover all privacy and security needs, particularly when this is overlooked during an emergency. There are difficulties to creating universal policies that can address every issue (Chang, 2021). Vendor agreements/terms may gain unlawful ownership or dissemination of information via loopholes in FERPA compliance (Paris et al., 2021). The key is to move beyond FERPA compliance and understand the unique security and privacy needs of an institution's student body.

Issues of privacy violations exist outside of data interruption and vague vendor licenses. Instructors may choose to use social media tools in their courses as a way to foster a class that is more engaging and attractive. Students' familiarity with social media makes integrating the tool into coursework simple. This is helpful during emergency remote learning when students are forced into a modality they did not originally choose. Nonetheless, not all students will feel safe sharing details with instructors and classmates. The use of social media in a formal learning context can blur the lines between personal and professional (Dennen & Burner, 2017). Some students may not be willing to share, for educational purposes, details such as photos socializing or certain posts from family and friends. Allowing a professor and classmates to "follow" one's Facebook page for the duration of a course could be awkward and uncomfortable (Dennen & Burner, 2017). Students are often at the mercy of the course's requirements, and if the incorporation of social media is necessary for achievement in the course, then a student does not have a choice. Emergency remote learning removed many opportunities for students to have a say in their educational options. Instructors should be open to providing students with some agency within the course. This can encourage students to appreciate the level of control they possess despite the uncontrollable circumstances of a health crisis.

Risks with Online Tools

Ultimately, institutions must reflect on whether their delivery of emergency remote learning, which began in 2020 and continues in some form currently, complied with FERPA (Gelpi, 2020). Universities are encouraged to exercise care when choosing a tech vendor for educational purposes (Williams et al., 2019). However, the hurried response to COVID-19 forced the adoption of tools that were not originally created for educational purposes but used for the classroom. The widespread use of such tools by individual instructors did not allow for an institution to monitor the technology being implemented in each classroom. Moreover, institutions were unable to offer speedy resources to make this massive shift less stressful for instructors. Utilizing free online tools made online instruction bearable despite the risks to privacy and security. There are instances where personal data is routinely exposed while using an online tool. These include creating a user account and saving work to commercial cloud storage. Downloading browser extensions are sometimes required for an online product to function on a personal computer or laptop. This can open a device to malware which may compromise a user's data and the device itself (Varshney et al., 2018). Instructors and students using video conferencing tools allow their faces, voices, and full names to be shared and recorded without the knowledge of where a recording is stored or what happens to it after a semester concludes. Such activities, associated with free online tools, can risk exposing student and instructor data. Traditional educational technology tools are not exempt from this concern. Institutions must outline specifics when it comes to vendor agreements and terms of service to minimize the exploitation of their students and instructors. Instructors should take precautions and verify policies outlined on terms of service pages when choosing certain tools for classroom use.

Data Breaches & Lawsuits

Higher education institutions continuously faced data breaches and challenges prior to 2020. However, this increased as emergency remote learning became the standard during the COVID-19 pandemic. Attacks to data security and privacy may come in the form of covert, deliberate theft and misuse of data from free online tools. Google is relied upon by many but remains a major data privacy and security offender. The company tracks user searches to personalize the user's experience. Google collects data and provides a limited ability for registered users to prevent the sharing of their information (Goodson, 2012). Google's suite offers a host of services used by students and instructors for both personal and professional purposes. Its ability to surveil online activity allows it to successfully sell its audience preferences to advertisers (Kang & McAllister, 2011). This strategy threatens the privacy of its users and exposes sensitive information to outside vendors. In addition to this, policies adopted focus more on maintaining Google's rights to user information (Kang & McAllister, 2011). This has led to lawsuits and settlements by the company regarding user data. In 2016, lawsuits were filed by the University of California system, specifically the Berkeley and Santa Cruz campuses, in which Google was accused of procuring student emails without consent from Google's Apps for Education suite (Riddell, 2016) and sharing them with advertisers. This case underscores the limited accountability and oversight to which Google is subjected. According to the lawsuit, Google scanned student information for years until the company formally announced that it "permanently removed all ad scanning" (Brown, 2016) from its education email service. Google continues to grow as a provider of technology in K-12 and post-secondary schools. The well-known browser Google Chrome is popular and used regularly by students and instructors. However, Google Chrome is no stranger to lawsuits that highlight the company's harvesting of data regardless of a user choosing to opt out (Nayak, 2022). As of today, Google declared a ban, beginning in late 2023, on allowing advertisers to track consumers in Chrome (Nayak 2022).

Hacking is another threat to student data. Since 2005, over 1,850 data breaches in education have been recorded nationwide, with 65% occurring in universities and colleges (Cook, 2021). Many of those breaches are the result of malware, spyware, and ransomware. All three attacks allow a third party to gain access to data (Grama, 2014) without consent and/or place data at high risk. This type of hacking can cause a myriad of serious problems for users. Stored data becomes compromised by exploiting weaknesses such as poor passwords, unsecured networks, or negligent security protocols (Beaudin, 2015). Whether the information is stored in a user account for an online tool or on an actual device, both locations are vulnerable once hacking has taken place.

Colleges and universities suffer tremendously when targeted by such mass data breaches. Institutions face monetary losses, suffer reputational consequences, and may experience reduced student enrollment (Grama, 2014). As a result of emergency remote learning during the COVID-19 pandemic, schools nationwide experienced a high number of breaches in 2020, with 2.99 million student records affected (Cook, 2021). The influx of students and instructors hurriedly working online placed more data in jeopardy. In 2020, Kansas City's Metropolitan Community College faced ransomware attacks often (Lubinski, n.d.) with the social security numbers, medical, and banking information of 630,000 former, current, and prospective students exposed (Cook, 2022). The well-known historically Black college and university (HBCU), Howard University, suffered ransomware attacks that locked its networks for days (Ngo, 2021). This forced online and hybrid classes to be canceled. The college continues to work with FBI and city officials regarding appropriate protection against cyberattacks. Unfortunately, some colleges are unable to recuperate their losses or recover altogether. Lincoln College, in Illinois, experienced a ransomware attack that took months to rectify (Holt, 2022). The COVID-19 pandemic caused severe reductions in student enrollment, recruitment, and fundraising. Consequences from the pandemic coupled with the ransomware attack overwhelmed the college. The 157-year-old institution could not survive and shut its doors on May 13, 2022 (Holt, 2022). Other institutions face security breaches from widely used online software. Stanford, the University of Colorado, the University of Miami, and the University of California system campuses Berkeley, Davis, and Los Angeles used the file-sharing program/service known as Accellion. In 2021, a data breach in the program resulted in the online publishing of student data from the aforementioned colleges (Wu & Catania, 2021).

Recommendations

Emergency remote learning should be safe. Now, in the third year of the pandemic, aspects of this type of learning continue to re-shape higher education moving forward. University students and instructors are slowly returning to a form of in-person teaching and learning. Simultaneously, a significant percentage of course offerings at various institutions may remain online for the foreseeable future. Other institutions might simply be interested in maintaining a robust online presence through growing their online courses. Since 2020, there have been trainings and guidelines developed to offer help to instructors when using specific tools adopted during the pandemic. For example, The City

University of New York (CUNY) (2020) created a website primarily for instructors providing Zoom guidance to ensure security and privacy. Regarding video participation, CUNY (n.d.) has organized a chart comparing the privacy options among the four tools used with video capability: Zoom, Teams, Blackboard, and Webex. This can be helpful for instructors when deciding which video conferencing tool is best to use for their students. Beyond the institutional level, there are handbooks for online course development. The Online Consortium (2020), in partnership with the Association of Public and Land-grant Universities and Every Learner Everywhere, created a faculty playbook. Its goal is to aid instructor readiness regarding online teaching. It provides information, resources, tips, and best practices, emphasizes equity, and identifies course design components to follow. It could serve as a much welcomed handbook for instructors still learning about effectively delivering courses in an online modality.

Free, easy-to-use tech tools for education have allowed instructors to stay connected with students, especially since the start of the COVID-19 pandemic. There is an emphasis on collaboration and sharing in an online environment, but this should not be at the cost of privacy (Chang, 2021). The protection of safe learning is ensured when the student's privacy can be guaranteed. Online course development trainings can help instructors build online courses with privacy and security in mind. Instructors can incorporate privacy criteria into the syllabus and create collaborative assignments that allow anonymizing information (Chang, 2021). Transparency helps in creating trust between the instructor and students; therefore, being open about the safety protocols to students should be encouraged. Students have a right to know how their information will be used, if at all, who will see it, why it is needed, where it will be stored, and how it will be destroyed (Vu et al., 2019). Instructors should be trained in FERPA policies and learn about policies beyond that. The U.S. Department of Education's (2014) "Protecting Privacy While Using Online Educational Services: Requirements and Best Practices" document should be updated, but it remains relevant. It covers information for instructors and institutions to follow for protecting student privacy. It reminds readers to be aware of state and institutional policies and encourages caution when using click-wrap consumer applications.

Since the start of the pandemic, institutions have more cause to negotiate beneficial agreements with third-party vendors. Even if a provider is designated as a "school official", the terms of service must be specific regarding the level of control over collected information (Gelpi, 2020). Contracts should be able to ensure that information collected is stripped of identifiers (Gelpi, 2020). An agreement may outline what data is permitted for collection and compensation for the institution if the terms are breached. The ownership of personal and identifiable data should be defined with protocols in place that do not leave student information exposed when a contract is terminated (Gelpi, 2020).

Agreements should include language on the destruction of student data upon the termination of a contract. At a very basic level, FERPA compliance (Paris et al., 2021) should be followed by an agreement with third-party vendors. Institutions can help instructors by periodically providing FERPA workshops and making available contracts with educational technology vendors for instructors to view. Faculty will continue to utilize easy-to-use free online tools for educational purposes. Rather than enforcing certain tools over another, institutions can create a list of teaching tools their faculty currently use, provide terms of service information for each, and offer a rating system designed to describe how well the tool indicates its protections of data and privacy. This will allow faculty to access terms of service information that might be difficult to find for some online tools. Institutions can encourage students and instructors to become involved in conversations about their privacy and security concerns, particularly in an emergency remote learning environment. Engaging both parties will allow institutions to notice the gaps in their local policies for safety in online learning.

Incorporating Multiple Voices: Collaborative Policy Building

Creating safe learning policies in response to emergency remote learning is best conducted collaboratively. There are many individuals within higher education that should have a voice in what they would like protected. Generally, instructors, staff, and students remain in the dark regarding the privacy and security of their data. It is important for all groups affected to come together and decide what kind of privacy and security issues should be addressed beyond that which FERPA covers. Four groups are identified below as stakeholders in creating privacy and safety policies. Collaborative policy building can begin with conversations surrounding the topic. The questions below are designed to help facilitate a discussion on safety and privacy in an online environment.

1. Teaching faculty, instruction librarians, and instructional staff is a group of people who do not have the title of faculty but engage with students and create content for teaching and learning in the classroom: As facilitators and designers of courses, instructors can be made aware of the terms of service or agreements for specific tools and assess whether they are best to use or how to use them to ensure safe online learning.

    a. What matters to you most in the privacy and security of education data that is personal, sensitive, and identifiable?
    b. What do you know about FERPA? What do you know about the institution's data security and privacy guidelines?
    c. How do you design your course with these principles in mind?

d. What are the educational technology tools that you use for your classroom?

2. Students comprise the group that is meant to be protected; therefore, including them in creating policies for online classroom privacy and safety is imperative. Students may trust the institution and instructor are making the right decisions and following FERPA policies when using third-party vendors and free online tools. Regardless of this trust, an understanding of what to identify as most important in an agreement can be helpful.

    a. Have you been made aware of FERPA or the institution's guidelines regarding your rights in the classroom?
    b. What are some educational tools that you have used in your classes?
    c. What are some of your privacy and security concerns with those tools and with learning online?
    d. How can the university help bring awareness to students about privacy protections and safe online learning?

3. Information technology department staff, specifically those in charge of educational technology tools. This group is in the best position to evaluate the privacy and security needs required of technology tools in the classroom. This group is also well-equipped to train students and instructors on understanding the importance of privacy and security protocols, FERPA guidelines, and other safeguards in online learning.

    a. What are some important factors instructors and students should identify when reviewing the terms of service for a free online tool/application?
    b. How might education for instructors and students on data privacy and security impact the college community?
    c. What is this department's role, if any, in choosing educational tools for campus-wide use?
    d. How does your department investigate and resolve data breaches or other data security issues?

4. University/College Administrators control the decision-making for the institution. This includes decisions on educational technology tools used campuswide. Administrators can support the privacy and safety concerns of instructors and students while working with IT department representatives to create proper safety policies for the institution. Individuals in this group have access to university funds and create the campus-wide budget. Funding can be expanded to include the purchasing of effective safety mechanisms for online teaching tools used by instructors.

a. Describe the process, from beginning to end, of acquiring educational tech tools for instructor use campus-wide.
b. What are the factors that are involved in deciding to adopt an educational tool?
c. How do information privacy and safety play a role in these decisions?
d. Does the college work with outside local or state cybersecurity divisions? How would this collaboration benefit or hinder the protection of student data?
e. What campus-funded online teaching and learning professional development programs exist for instructors?

Conclusion

The scramble to place material online and continue classroom activities became a priority, with no time to consider data privacy and security protection. Universities and colleges focused on maintaining educational services for students throughout the beginning of the pandemic. Overnight, instructors were placed in challenging situations, without guidance on how to transfer all in-person classes to fully online. A few years into the pandemic and instructors have improved their online teaching, become more familiar with distance learning pedagogy, and have more resources, supported by the administration, to aid in online course development. It is time to review vendor contracts hastily accepted and investigate the terms of service for free online tools and social media adopted during the pandemic. This is a great moment for institutions to explore existing and new data privacy and security issues that have arisen with the majority of classes taking place online. Administrators, IT staff, instructors, and students have an opportunity to come together and create local policies that directly address privacy and safety. Online learning for traditionally face-to-face universities and colleges will continue while the pandemic endures. For this reason, it is important to tackle and share knowledge about data protections for students to ensure emergency remote learning is successful and safe.

References

Ali, M. B. (2021). Multi perspectives of cloud computing service adoption quality and risks in higher education. In M. Khosrow-Pour (Ed.), Handbook of research on modern educational technologies, applications, and management (pp. 1-19). Information Resources Management Association. http://doi.org/10.4018/978-1-7998-3476-2

Beaudin, K. (2015). College and university data breaches: Regulating higher education cybersecurity under state and federal law. *Journal of College and University Law, 41*(3), 657.

Brooks, K. J. (2020, July 17). Zoom says it will fix security holes that video hackers have exploited. CBS News. Retrieved August 22, 2022, from https://www.cbsnews.com/news/zoom-video-conferencing-feature-freeze-security-flaws/

Brown, E. (2021, October 27). UC-Berkeley students sue Google, alleging their emails were illegally scanned. *The Washington Post*. Retrieved August 22, 2022, from https://www.washingtonpost.com/news/grade-point/wp/2016/02/01/uc-berkeley-students-sue-google-alleging-their-emails-were-illegally-scanned/

Center for Disease Control and Prevention. (2018). Family Educational Rights and Privacy Act (FERPA). https://www.cdc.gov/phlp/publications/topic/ferpa.html

Center for Disease Control and Prevention. (2022). CDC museum COVID-19 timeline. https://www.cdc.gov/museum/timeline/covid19.html

Chang, B. (2021). Student privacy issues in online learning environments. *Distance Education, 42*(1), 55–69. https://doi.org/10.1080/01587919.2020.1869527

City University of New York. (2020). Zoom security protocol. https://www.cuny.edu/wp-content/uploads/sites/4/page-assets/about/administration/offices/cis/it-resources-for-remote-work-teaching/Zoom-Security-Protocol.pdf

City University of New York. (n.d.). Video participation privacy options. Remote Learning & Work. https://www.cuny.edu/wp-content/uploads/sites/4/page-assets/about/administration/offices/cis/it-resources-for-remote-work-teaching/Video-Participation-Options.pdf

Cook, S. (2022, January 21). US schools leaked 28.6 million records in 1,851 data breaches since 2005. Comparitech. Retrieved August 22, 2022, from https://www.comparitech.com/blog/vpn-privacy/us-schools-data-breaches/

Dennen, V. P., & Burner, K. J. (2017). Identity, context collapse, and Facebook use in higher education: Putting presence and privacy at odds. *Distance Education, 38*(2), 173–192. https://doi.org/10.1080/01587919.2017.1322453

Elmer, G., Neville, S. J., Burton, A., & Ward-Kimola, S. (2021). Zoombombing during a global pandemic. *Social Media + Society, 7*(3), 1–12. https://doi.org/10.1177/20563051211035356

Gelpi, A. (2020). Ensure FERPA compliance in online provider agreements. *Campus Legal Advisor, 20*(11), 1–5. https://doi.org/10.1002/cala.40272

Grama, J. (2014). Just in time research: Data breaches in higher education. EDUCAUSE. https://library.educause.edu/resources/2014/5/just-in-time-research-data-breaches-in-higher-education

Goodson, S. (2022, April 14). If you're not paying for it, you become the product. Forbes. Retrieved August 22, 2022, from https://www.forbes.com/sites/marketshare/2012/03/05/if-youre-not-paying-for-it-you-become-the-product/

Hodges, C. B., Moore, S., Lockee, B. B., Trust, T., & Bond, M. A. (2020). The difference between emergency remote teaching and online learning. EDUCAUSE. https://er.educause.edu/articles/2020/3/the-difference-between-emergency-remote-teaching-and-online-learning

Holt, K. (2022, May 9). A US college is shutting down for good following a ransomware attack. Engadget. Retrieved August 22, 2022, from https://www.engadget.com/lincoln-college-ransomware-attack-shut-down-covid-19-164917483.html

Ibacache, K., Rybin Knoob, A., & Vance, E. (2021). Emergency remote library instruction and tech tools. *Information Technology and Libraries, 40*(2), 1–30. https://doi.org/10.6017/ital.v40i2.12751

Kang, H., & McAllister, M. P. (2011). Selling you and your clicks: Examining the audience commodification of Google. *tripleC: Communication, Capitalism & Critique, 9*(2), 141–153. https://doi.org/10.31269/triplec.v9i2.255

Karakaya, K. (2020). Design considerations in emergency remote teaching during the COVID-19 pandemic: A human-centered approach. Educational Technology Research and Development, 69(1), 295–299. https://doi.org/10.1007/s11423-020-09884-0

Kim, S. S. (2021). Motivators and concerns for real-time online classes: Focused on the security and privacy issues. *Interactive Learning Environments*. https://doi.org/10.1080/10494820.2020.1863232

Lubinski, A. (n.d.). MCC Kansas City victim of cyberattack. *Courier Tribune*. Retrieved August 23, 2022, from https://www.mycouriertribune.com/schools/higher_education/mcc-kansas-city-victim-of-cyberattack/article_cf9b29f2-de35-11ea-9cd8-4312b8110d16.html

Nayak, M. (2022, February 28). All the ways Google is coming under fire over privacy: Quicktake. Bloomberg. Retrieved August 22, 2022, from https://www.bloomberg.com/news/articles/2022-02-28/all-the-ways-google-is-coming-under-fire-over-privacy-quicktake

Ngo, M. (2021, September 7). Howard University hit by a ransomware attack. *The New York Times*. Retrieved August 22, 2022, from https://www.nytimes.com/2021/09/07/education/howard-university-ransomware.html

Nørgård, R. T. (2021). Theorising hybrid lifelong learning. *British Journal of Educational Technology, 52*(4), 1709–1723. https://doi.org/10.1111/bjet.13121

Paris, B., Reynolds, R., & McGowan, C. (2021). Sins of omission: Critical informatics perspectives on privacy in e-learning systems in higher education. *Journal of the Association for Information Science and Technology, 73*(5), 708–725. https://doi.org/10.1002/asi.24575

Riddell, R. (2016, May 18). Google sued by 890 students over unlawful data mining allegations. Higher Ed Dive. Retrieved August 22, 2022, from https://www.highereddive.com/news/google-sued-by-890-students-over-unlawful-data-mining-allegations/419368/

Shearer, R., Aldemir, T., Hitchcock, J., Resig, J., Driver, J., & Kohler, M. (2020). What students want: A vision of a future online learning experience grounded in distance education theory. *American Journal of Distance Education, 34*(1), 36–52. https://doi.org/10.1080/08923647.2019.1706019

Tarisayi, K. S., & Munyaradzi, E. (2021). A simple solution adopted during the Covid-19 pandemic: Using WhatsApp at a university in Zimbabwe. *Issues in Educational Research, 31*(2), 644–659.

U.S. Department of Education. (2014). Protecting student privacy while using online educational services: Requirements and best practice. https://tech.ed.gov/wp-content/uploads/2014/09/Student-Privacy-and-Online-Educational-Services-February-2014.pdf

U.S. Department of Education. (2021). Family Educational Rights and Privacy Act (FERPA). https://www2.ed.gov/policy/gen/guid/fpco/ferpa/index.html

Varshney, G., Bagade, S., & Sinha, S. (2018). Malicious browser extensions: A growing threat: A case study on Google Chrome: Ongoing work in progress. 2018 International Conference on Information Networking (ICOIN), 188–193. https://doi.org/10.1109/ICOIN.2018.8343108

Vu, P., Adkins, M., & Henderson, S. (2019). Aware, but don't really care: Student perspectives on privacy and data collection in online courses. *Journal of Open, Flexible and Distance Learning, 23*(2), 42–51.

Walsh, L. L., Arango-Caro, S., Wester, E. R., & Callis-Duehl, K. (2021). Training faculty as an institutional response to COVID-19 emergency remote teaching supported by data. *CBE–Life Sciences Education, 20*(3), ar34. https://doi.org/10.1187/cbe.20-12-0277

Williams, D., Kilburn, A., Kilburn, B., & Hammond, K. (2019). Student privacy: A key piece of the online student satisfaction puzzle. *Journal of Higher Education Theory and Practice, 19*(4), 115–120. https://doi.org/10.33423/jhetp.v19i4.2206

Wu, D. & Catania, S. (2021, April 3). Hackers leak Social Security numbers, student data in massive data breach. *The Stanford Daily*. Retrieved August 22, 2022, from https://stanforddaily.com/2021/04/01/hackers-leak-social-security-numbers-student-data-in-massive-data-breach/

"At the Cost of My Well-being": Exploring Trans, Non-binary, and Gender-Diverse Students' Experiences of Online Learning

Maddie Brockbank, Wil Fujarczuk, Christian Barborini, and Yimeng Wang

This project emerged as a response to the unique experiences of online learning expressed by trans, non-binary, and gender-diverse students during the 2020-2021 academic year. By saying "trans" throughout this chapter, we recognize all transgender, non-binary, gender-diverse, Two-Spirit, and questioning folks. We also recognize that some who identify with labels mentioned above may not personally identify as trans and encourage readers to consult with individuals one-on-one to determine what labels best meet their preferences.

Like many academic institutions, the 2020-2021 academic year at McMaster University was the institution's first-ever predominantly online educational experience for the majority of students and instructors, due to mass shutdowns facilitated by the COVID-19 pandemic. While the widespread implementation of this form of education has generated significant discussion on pedagogical efficacy, largely missing from the conversation has been how online attendance and participation in the classroom affect trans students specifically. Throughout the year, many students shared reflections on both positive and negative aspects of their educational experience with undergraduate peer support volunteers at the Women and Gender Equity Network (WGEN) and the Pride Community Centre (PCC). These conversations frequently discussed the challenges of feeling unsafe in online learning spaces. As online learning continues to be of significance at McMaster University and in higher education more broadly, this area of research is a timely response to an emerging issue. Additionally, it maintains transferable links to and implications for blended and in-person learning as well.

In 2021, our team published the findings of our study in a short report for instructional purposes. The original abridged report served as a tangible, albeit introductory, resource for teaching teams and university administration regarding facilitating safe(r) spaces for gender-diverse students in online, blended, and in-person learning spaces. The resource is informed by students' experiences and narratives expressed during their participation in the study regarding their educational experience over the past year. As a main value held by the co-investigators, the resource sought to reflect and attend to concerns identified specifically by trans, non-binary, and gender-diverse students at McMaster University.

This chapter aims to provide a deeper analysis into the experiences and recommendations posed by trans students around remote learning and the eventual transition into in-person or blended instruction. The structure of this chapter is as follows: (1) we begin with reviewing relevant literature and its gaps in articulating trans students' experiences in the postsecondary context; (2) we then draw on minority stress theory, qualitative data collection procedures, and thematic analysis in generating and analyzing our data; (3) we outline central themes identified by our team about trans students' experiences of course instruction, policy, and practice; (4) we link these findings to the larger themes articulated in burgeoning literature; (5) we pose point-form recommendations for educational stakeholders regarding building trans-inclusive classrooms; and (6) we conclude with a recognition of limitations and a brief discussion to implications for future work in this area.

Literature Review

Emerging literature has begun to explore the experiences of trans and non-binary students in a postsecondary context, specifically in examining the unique barriers these students face that impact their ability to fully engage in their education (Beemyn, 2005; Goldberg, 2018; Goldberg & Kuvalanka, 2019; McLemore, 2018; Rankin, 2006; Schneider, 2010; Seelman, 2014; Siegel, 2019; Swanbrow Becker et al., 2017; Whitley et al., 2022). In particular, research has indicated that trans students experience specific barriers to safety, inclusion, and representation in the academy, which then shape their feelings of safety and belonging both in the classroom and on campus. For example, Goldberg's (2018) report documenting the experiences of trans and non-binary postsecondary students indicates that they are subjected to significant levels of discrimination and harassment from their peers, instructors, and administrators, thus influencing their perceptions of hostile campus and classrooms climates. Further studies build on these findings to demonstrate the ways in which hostile campus climates facilitate trans students' concerns regarding physical and mental health, distress levels, sense of community, and academic performance during their postsecondary education (Beemyn, 2005; Goldberg, 2018; Rankin, 2006; Swanbrow Becker et al., 2017).

While marginalized students in general have been shown to experience discrimination that shapes their postsecondary engagement, trans and non-binary students report specific instances of harm that are unique to trans people (Goldberg, 2018; Goldberg & Kuvalanka, 2019; Seelman, 2014; Siegel, 2019; Whitley et al., 2022). Namely, misgendering, deadnaming, and outing have been identified by trans students as some of the most significant forms of interpersonal or microlevel transphobia that they experience during their education (Faris, 2019; Goldberg, 2018; McLemore, 2018; Sinclair-Palm & Chokly, 2019; Whitley et al., 2022). Misgendering is a term used to describe interactions where a trans student is referred to in a way that misaligns with or contradicts their gender identity (McLemore, 2018; Whitley et al., 2022), such as by using "he/him" pronouns for someone who uses "she/her" pronouns. Misgendering can occur intentionally or unintentionally and is often linked to the preconceived notions of how a person of a certain gender "should" look (e.g. facial hair, vocal register, body shape, clothing). Deadnaming refers to the act of calling a trans person by their birth name or other former name, either intentionally or unintentionally, which invalidates a person's identity (Goldberg, 2018; Sinclair-Palm & Chokly, 2019). Outing is a practice of revealing someone's sex assigned at birth, gender, or sexual orientation without their consent (Pryor, 2015).

While many of these instances of harm appear on an interpersonal level, they point to the systemic, institutional, and structural nature of marginalization and violence, whereby cisheterosexism and transphobia are foundational to Western academia and built into the very practices of postsecondary education (Formby, 2017; Fraser, 2020; Maughan et al., 2022; Siegel, 2019). For example, Western education is still largely dictated by binary perspectives of gender that rely on alleged biological and sex-based differences (Eldridge, 2020; McPhail et al., 2016). Further, it is significant to understand the relationship between transphobia and other systems that facilitate marginalization and violence, including racism, colonialism, and ableism. Expanding on the work of Lugones (2016), who wrote about the colonial origins of the gender binary, emerging work has explored the relationship between ongoing practices of colonization, eugenics, and imperialism and how they reify and perpetuate dichotomous, Westernized constructs of gender, sex, and identity (Ballestín, 2018; Kravitz, 2020; Omowale, 2021). A failure to apply a historiographical analysis to curriculum that discusses gender can facilitate the dynamics mentioned above, thus (re)entrenching cisheterosexist discourse shaping how people understand trans people and their experiences.

Massive shifts in postsecondary education facilitated by the COVID-19 pandemic have exposed a gap in understanding the experiences of trans students in remote learning. Prior to and since the large-scale shutdowns of in-person learning and, consequently, the rather abrupt shift to remote course instruction, burgeoning literature has begun articulating the unique experiences of marginalized students in online education (James, 2021; Kimble-Hill et al., 2020; Phirangee & Malec, 2017). However, despite some newer literature (e.g. Gonzales et al., 2020; Mavhandu-Mudzusi et al., 2021; Whitley et al., 2022), there continues to be a limited understanding of how the pandemic specifically impacted trans students' ability to engage in their learning.

Theory/Methodology/Methods

This study incorporates the minority stress theory framework to conceptualize how the experiences of trans students translate to the health disparities observed within this population (Hendricks & Testa, 2012; Meyer, 2003). Minority stress, which exists within the realm of social stress theory, refers to the excess stress experienced by marginalized populations because of their social position (Meyer, 2003). Minority stress theory posits that social conditions, such as prejudice, stigma, and discrimination, foster a hostile and stressful social environment, which amounts to mental and physical health challenges. We draw on minority stress theory to illustrate the role of stigma, discrimination, and prejudice in the amplification of challenges that affect the physical, mental, and social health of trans students who have already experienced a variety of obstacles throughout the COVID-19 pandemic.

In June 2021, we received ethics clearance to conduct an online qualitative survey and online follow-up interviews, which were advertised predominantly through social media. We provided students with a letter of information prior to taking the survey and/or participating in the follow-up interview. Students who participated in the follow-up interviews were assigned pseudonyms to identify them. Since the surveys were submitted anonymously, no pseudonyms were assigned; however, we draw on different survey responses throughout this findings section to represent the responses of all those who participated in the study.

The questions on both the survey and in the interview focused on participants' experiences of transitioning to online learning, the benefits and drawbacks of remote learning, and tangible recommendations for how online learning could be improved to better support trans students. A total of 15 questions were included in the survey (see appendix A) that explored students' experiences of online learning. The follow-up interview guide included six prepared questions (see appendix B) and were semi-structured in nature to provide students with the opportunity to further elaborate on their responses to the survey questions. Our online survey garnered 22 full responses from trans students, while seven participated in follow-up interviews on Zoom.

Following data collection, our team engaged in thematic analysis of the interview transcripts and written survey responses. Thematic analysis (TA) is "a [descriptive] method of identifying, analyzing, and reporting patterns (themes) within data" (Castleberry & Nolen, 2018, p. 808), which we drew upon due to its wide applicability and flexibility in creating space for literal, interpretive, and reflexive readings of the data. TA seeks to identify themes related to the research question(s) and (re)articulate the purpose of the study (Braun & Clarke, 2014; Castleberry & Nolen, 2018). We followed Castleberry and Nolen's (2018) steps for thematic analysis, which include compiling (transcribing and organizing the data), disassembling (coding the data), reassembling (identifying themes within the codes), interpreting (making analytical conclusions about the themes identified), and concluding (positioning these themes in relation to the research question).

Findings

Our team developed the following themes in a shared thematic analysis. After identifying these themes, we sent our preliminary analysis to participants who indicated their interest in reviewing them for feedback before proceeding with writing the report. These themes are situated broadly within educational, administrative, physical, and institutional barriers that trans students identified as impacting their education. However, we have broken these broader categories into a brief series of themes that capture the essence of the harms that trans students in our study identified as most salient.

Expectations of Pronoun Disclosure

Depending on the remote learning platform used, participants on a video call can add or change their name and pronouns to be visible on the screen. For example, while Zoom has the function to change your screen name, Microsoft Teams does not; rather, Microsoft Teams uses the name attached to a student's account (e.g. usually a legal name) and requires administrative permission to edit. Many participants in our study identified the challenges of these platforms, where the online learning technology itself might facilitate deadnaming and misgendering.

Further, participants identified common practices that instructors viewed as fostering inclusivity as facilitators for experiences of transphobia, including outing, deadnaming, and misgendering. As one participant summarized in a survey response:

> "There is an increased pressure to display pronouns by professors—even though it is not safe for me to do so as I am not out publicly yet. While I think it is important for professors to make it a suggestion, I have had a few professors call people out by name and almost force them into doing it. Forcing me to lie and display pronouns that do not fit me in order to feel safe in the class setting."

Here, instructors' attempts to normalize pronoun disclosure led to feelings of discomfort among students who continue to navigate and negotiate their identity personally and publicly. In a follow-up interview, Alex added:

> "People really push the whole 'pronouns in their name,' they really push that on Zoom or they really push it in, like, when you introduce yourself, or whatever. And I really value that, and I think that it creates a really important space of a safer, braver kind of space for people. But as someone who would rather just not address that and as someone who would rather not talk about pronouns because I don't know what the answer is and I'm okay with not having a concrete answer."

These experiences were observed across both interview and survey responses in our data set. While these practices can certainly build opportunities for safer spaces that value respecting pronouns, they risk presenting pronouns as a static practice (e.g. a singular set of pronouns). Through this process, trans students who might have multiple pronouns, whose pronouns change temporally or contextually, and/or who do not feel safe giving their pronouns risk being outed or called upon by their instructors for not following this informal course policy.

Demanding "Cameras On"

In the transition to online learning, many instructors developed informal course policies around evaluating students' participation, which may have included requiring that students keep their cameras on during class so that they are visible on the screen. Some students spoke positively of class policies that did not require cameras on. For example, as one student wrote in response to a question about the benefits of online learning included in the survey:

> "I feel I have more control over how I express my gender. For example, I can turn my camera off if I'm feeling dysphoric, I can include my pronouns in my Zoom name, etc. Behind a screen, I have less worry about how I present myself on camera and have the option to keep my camera off entirely."

The ability to selectively and intentionally present or make themselves visible was viewed largely positively by students in our study. When many trans people experience gender dysphoria, which here refers to significant distress regarding the expression and perception of one's gender, creating opportunities for them to exercise agency in how they engage physically in a space can facilitate safety. Further, when someone might be participating in remote learning in a location that is not safe for them to express their gender (e.g. unsafe or transphobic home or public environment), having the option to participate without cameras on might mitigate experiences of dysphoria.

However, students in our study also spoke about how many of their classes required cameras on for participation grades. As one student notes in their survey response,

> "Some of my classes, professors, TAs did not require students to have their cameras on during lectures/tutorials. However, some of them did have this requirement, which was tied to a portion of one's grade in the course, and as a non-binary student with very strong experiences of dysphoria, it made me feel really awful and prevented me from fully participating in a course's lectures when I was forced to have my camera on. Seeing myself on a screen in a way over which I have little to no control was very triggering to my dysphoria and, as a result, my mental health. Some professors didn't seem to understand that seeing yourself on a Zoom call is different than seeing yourself in a selfie you post online—in the latter, I am in control of how I look and who sees me. In the former, that control is taken away from me and I am forced to comply for the sake of a grade, at the cost of my wellbeing."

Here, this participant articulated a shared theme among responses and interviews: that mandatory "cameras on" policies do not consider the discomfort, distress, and fear that many trans students experience due to feeling a loss of control over how they present themselves and are perceived by others. As a result, students reported a disengagement from their learning (e.g. not feeling like they could fully or actively participate in their classes) and significant experiences of distress (e.g. mental health concerns).

This dynamic is exacerbated by expectations of how a student should participate in a remote class, including raising their hand, unmuting their microphone, and speaking during a video call (some of which are recorded and uploaded to course sites). Students in our study spoke of dysphoria extending beyond visual presentation/perception to vocal registers and how their voice might be received as incongruent with their gender expression.

Different Disciplines

While experiences of transphobia were significant across faculties and departments in the university, as indicated by participants' survey responses and the myriad of disciplines represented in our data, there were distinctions between and among disciplines. Namely, students in our study indicated that certain departments were seen as more accepting, welcoming, and committed to trans-inclusivity than others, which they suggested was an unspoken yet shared understanding among trans students at McMaster University. As Devon stated during a follow-up interview:

> "Cultural studies is where you get queer studies, where you get trans studies, mad studies, all that stuff that is going to potentially draw the people who are affected by those issues and who live those issues, who are the ones who are studying them. And then when I think about other departments – I mean, I haven't taken a ton outside of the Humanities, but I've taken a little bit in Social Sciences and Economics, and those are, again, I think you're seeing less representation of these different marginalized groups in the faculty level."

Devon's comment was further reflected in other survey and interview responses, which focuses on the ways in which disciplines emphasizing the significance of lived experience in their curriculum and in faculty representation tend to be more equipped to discuss issues specific to trans communities. Many students indicated that their department had no out trans or gender-diverse faculty, which made it difficult to connect with instructors and teaching teams about their concerns. Other participants spoke of the differences between Humanities and Social Sciences as compared to Health and Life Sciences, where the latter faculties were perceived as heavily reliant on binary constructs of gender that facilitated transphobic discourse in the classroom.

Intersectionality

A particularly salient theme emerging from our conversations with trans students highlighted the importance of applying an intersectional analysis to discussions of gender and understanding trans students' experiences. As Sam stated in a follow-up interview:

> "I think more of a conversation about intersectionality is important [...] if they are going to make it a space to show up for trans and gender-diverse people, I think it's particularly important for them to make space for other people's identities as well, like disabled people, people of colour, things like that. Because I know from personal experience and I know from listening to friends that you can't just show up for one part of your students' lives, and not any others."

Sam's comment reflected some shared concerns that students participating in our study emphasized in both survey and interview data. Here, students emphasized the need for instructors and courses to recognize and attend to the diverse experiences of their students along the axes of gender, race, and disability, among other identities, more readily. Another student spoke of this in specific relation to the need for a historiographical analysis of concepts taught in courses that point to their often white supremacist, cisheterosexist, and colonial origins, such as "IQ testing," eugenics, "sex-based or biological differences," and how many health interventions (particularly in the medical and psychological fields) universalize a White, cisgender, living without a disability, heterosexual man's experience. Applying an intersectional lens to these discussions would also make space for trans students who are not White or living without a disability to better understand their experiences as they are historically situated in broader projects of colonialism, imperialism, and ableism. As one student emphasized, understanding trans students' experiences cannot be done through a White, Western lens that only examines gender; rather, the confluence of these systems of oppression must be examined as mere cogs operating in a broader machine.

Discussion

Our study addressed a gap in current literature on the topic of trans students' unique experiences in remote learning. Particularly, by narrativizing students' experiences and situating it within the broader context of both (1) transitioning back into in-person learning and (2) posing action-oriented recommendations that are informed by trans students themselves, we recognize the depth and complexity of the topic, while also centering tangible solutions that educators can prioritize when designing their courses moving forward. Many students in our study commented on how much they appreciated being given the space to share their experiences, and they emphasized their sincere hope that these conversations resulted in positive changes for themselves and for generations of students to come.

Our first two themes, which discussed the drawbacks of newly created course policies that outline expectations for participation and engagement in remote learning, contextualize our use of minority stress theory, whereby many of our participants reported adverse mental health outcomes associated with distress experienced in online learning environments (Hendricks & Testa, 2012; Meyer, 2003; Whitley et al., 2022). These experiences also mirror what has been reported previously in literature on this subject, particularly as it relates to trans students' unique experiences of discrimination and their impact on distress levels, sense of belonging, and perceived safety in the postsecondary context (Beemyn, 2005; Goldberg, 2018; Rankin, 2006 Swanbrow Becker et al., 2017). While our survey only garnered 22 responses and, therefore, cannot be generalized, participants tended to report feelings of distress, discomfort, and fear on scaled questions when comparing their experiences on different online learning platforms to their experiences of in-person learning. These were further discussed via long-answer responses.

Participants' distress around pronouns disclosure and "cameras on" policies was further complicated by the complex contexts that many were also navigating while accessing remote learning. Specifically, many students engaging in remote learning may be doing so in home environments or public spaces where they are not out or are not respected by those in their physical proximity (e.g. family, friends, roommates, peers in a public space). It is also important to note that some students in our study discussed how campus policies demanding or requesting pronoun disclosure did not actually facilitate safer spaces; in reality, many students in our study discussed how their pronouns were not used by peers, regardless of how students identified. For example, one student spoke about sharing their pronouns and then consistently being misgendered, without correction, throughout the term as people relied on aesthetic assumptions/cues about the student's gender. One participant aptly summarized that pronoun disclosure often seemed like a perfunctory "checkbox," whereby facilitators or instructors asked for pronouns without considering the broader practices that are necessary for building trans-inclusive classroom spaces.

Further, framing specific kinds of engagement as "mandatory" for evaluative purposes risk exacerbating the very stressors that instructors might be trying to alleviate. For example, when students fear being penalized for not contributing to class discussions, policies around visual (cameras on) and verbal (microphones on) expectations of participation can facilitate discomfort for trans students and practices of misgendering, deadnaming, and outing (e.g. if someone's voice/Zoom image is used to deduce gender/pronouns rather than what the student indicates their gender/pronouns are). These experiences mirror those identified in the literature around the unique barriers that trans students are subjected to when engaging in postsecondary study, whereby course structure might be built in a way that puts students at risk of hypervisibility, hostility, violence, and marginalization (Formby, 2017; Goldberg, 2018; Whitley et al., 2022).

The latter themes, discipline-specific experience and intersectionality, shift away from hyperfocus on interpersonal harm to reveal the institutional and systemic nature of many of the concerns discussed. For example, some disciplines are seen to be fundamentally cisheterosexist in nature based on their foundational assumptions/concepts about gender, "sex," biology, and identity. Further, those courses that do discuss gender often do so through a White, Eurocentric lens that reifies monolithic perspectives of gender and (re)centres White trans peoples' experiences. The concerns identified by participants in our study demonstrated significant critical analysis introduced by Lugones (2016) and others, who have written about the coloniality of the gender binary and the ways in which it has facilitated, sanitized, and justified other violent practices of transphobia, colonization, racism, and ableism (Ballestín, 2018; Kravitz, 2020; Omowale, 2021). Here, we see that many of these issues are embedded in the very fabric of Western academia, which demands mass upheaval of Western education as we know it.

Recommendations

Based on the experience of the students who participated in this study, we have drafted the following recommendations to create more trans-inclusive learning spaces. While these suggestions may assist educators in shifting classroom spaces toward safety, we acknowledge that many of these recommendations require institutional support and structural changes, some of which we briefly discuss in the final section of this report. Additionally, as mentioned in the previous section, we also know that many structural changes would require significant upheaval. With that in mind, we situate these recommendations in the broader awareness that they require institutional support and larger advocacy efforts on the part of educational stakeholders.

We have organized the following five recommendations in point form for readability and accessibility purposes. In our experience, educators are compelled to engage in bite-sized, straight-forward, and tangible recommendations that are situated in broader analyses.

Self-Education

- Familiarize yourself with resources for trans students at your school and in your community
    - Is gender-affirming counselling or healthcare available?
    - What gender-affirming peer support services exist?
    - Which department oversees name change processes for students, and are the staff there gender-affirming?
    - Is there local legal support for name change and gender marker change on government ID?
- Seek out workshops run by trans and gender-diverse people who discuss the nuances of trans identity, pronouns, and trans-specific issues; ensure those running the workshop are fully compensated for their labour
- Learn about the ways trans identities intersect with other identities (e.g., disability, race)
- Listen to how trans and gender-diverse students ask to be supported

Pronouns

- Normalize introducing pronouns by including your own where relevant
    - Introductions (e.g., "Hello class, my name is Dr. Malik, I use he/him pronouns")
    - On screen name (e.g., "Dr. Malik (he/him)")

- - - Some virtual platforms have created pronouns sections so they are automatically included where your name appear
    - If a pronouns section is not available, you can add your pronouns to your last name on your profile
  - In email signature (e.g., "Dr. Malik, PhD (he/him)")
- Create space for students to introduce their pronouns without making it mandatory
  - E.g., if going around and asking students to introduce themselves, you could let them know that they can do so by including "one or all of the following: name, pronouns, year, program"
- Create space for students to share their preferred name; make sure you use this name, even if the online platform does not allow changes to be made or if it does not match the name listed on the class list
  - For smaller classes, you could create a "get to know you" form that includes preferred name and pronouns to be used in front of the class
- Use gender-neutral language when referring to the class and normalize this language when talking about issues, populations, topics, etc.
  - "Students," "scholars," "everyone," and "y'all" are preferred over "ladies and gentlemen"
  - Use "they" in writing rather than "he/she"
  - Encourage students to use gender-neutral (i.e., they/them) pronouns or the student's name when referring to classmates unless their pronouns were otherwise specified

Facilitating Synchronous Sessions

- Avoid mandating that students keep their cameras on
  - If you would prefer students have their cameras to foster a sense of connection, especially for smaller classes, use language such as "Cameras on is encouraged, though not mandatory"
  - Tell students they can turn their cameras on or off through the session as needed
- Provide multiple format options for students to engage sessions to accommodate those uncomfortable using the microphone
  - E.g., "In the chat box, or by raising your hand, what are your thoughts?"
  - Consider the physical barriers that students may be experiencing, including where they are while attending virtual classes (e.g., they might be in a home environment where they

are not out), and accommodate these experiences as you learn about them

Administrative Considerations

- Include a section in your syllabus about equity and inclusion, informing students that discrimination and harassment based on gender identity and gender expression is prohibited
    - Include links to relevant school policies and offices
- Review course, department, faculty, and school-wide administrative documents to ensure trans-inclusive language is used, gender diversity is recognized, and students are given options to disclose
- Where possible, avoid the use of platforms that do not allow for changes to names or the inclusion of pronouns
    - If this is not possible, ensure that you make space for students to share their preferred names and pronouns (if they wish) and ensure you use these despite what is displayed on-screen
- Advocate to your department, faculty, and/or the school to ensure administrative tools and documents are more inclusive and respectful of trans identities

Institutional Considerations

- Offer students opportunities to shape how they can engage in the classroom
- Listen and compensate trans individuals for sharing ideas on how to make classroom spaces more trans-inclusive
- Hire more trans and gender-diverse instructors and staff
- Implement training for instructors on how to build more trans-inclusive spaces and create opportunities to practice implementing them
- Hold the institution accountable for the safety of trans students by making accountability measures for all instructors clearer

Limitations and Implications

This study was limited by a small sample size, where 22 students fully completed the survey and only seven participated in follow-up interviews. At one point during our study, we were forced to comb through our survey data when several hundred responses were submitted by 'bots' (fake survey responses). We also acknowledge that the study was impacted by COVID-19 and Zoom interviews; some participants had to reschedule interviews or adapt their participation due to being in unsafe home environments and fearing being overheard. Lastly, McMaster University has since articulated a commitment, like many other academic institutions, to return to primarily in-person learning, which renders some of what we have written irrelevant to in-person learning.

However, despite recognizing these limitations, our team firmly believes in the merit of our recommendations and their implications for online, blended, and in-person learning at postsecondary institutions across Canada and beyond. Education must be adapted to recognize the complex experiences of trans and other marginalized students in order to mitigate the barriers that shape their (in)ability to fully participate in their learning. As more is being written about building trans-inclusive classrooms (see De Pedro et al., 2016; Formby, 2017; Goldberg et al., 2018; Lawrence & McKendry, 2019; Seelman, 2014; Selander & Tidball, 2020), educational stakeholders must invest intentionally in creating safer and more accountable classroom and campus spaces.

References

Ballestín, L. (2018, July 26). *Gender as Colonial Object: The spread of Western gender categories through European colonization.* Public Seminar. https://publicseminar.org/2018/07/gender-as-colonial-object/

Beemyn, B. G. (2005). Making campuses more inclusive of transgender students. *Journal of Gay & Lesbian Issues in Education, 3*(1), 77–87. https://doi.org/10.1300/J367v03n01_08

Braun, V., & Clarke, V. (2014). What can "thematic analysis" offer health and wellbeing researchers?. *International Journal of Qualitative Studies on Health and Well-being, 9*(1), 26152. https://doi.org/10.3402%2Fqhw.v9.26152

Castleberry, A., & Nolen, A. (2018). Thematic analysis of qualitative research data: Is it as easy as it sounds? *Currents in Pharmacy Teaching and Learning, 10*(6), 807–815. https://doi.org/10.1016/j.cptl.2018.03.019

De Pedro, K. T., Jackson, C., Campbell, E., Gilley, J., & Ciarelli, B. (2016). Creating trans-inclusive schools: Introductory activities that enhance the critical consciousness of future educators. *International Journal of Teaching and Learning in Higher Education, 28*(2), 293–301.

Eldridge, J. J. (2020). Moving academia beyond the gender binary. *Research and Development in Higher Education: Next Generation, Higher Education: Challenges, Changes and Opportunities Volume 42*. Higher Education Research and Development Society of Australasia, Inc.

Faris, M. J. (2019). *On trans dignity, deadnaming, and misgendering: What queer theory rhetorics might teach us about sensitivity, pedagogy, and rhetoricity* [PowerPoint slides]. https://hdl.handle.net/2346/84247

Formby, E. (2017). How should we 'care' for LGBT+ students within higher education? *Pastoral Care in Education, 35*(3), 203–220. https://doi.org/10.1080/02643944.2017.1363811

Fraser, J. (2020). The struggle to imagine higher education otherwise: The transformative potential of diverse gender knowledges. *feminists@law, 10*(2). https://journals.kent.ac.uk/index.php/feministsatlaw/article/view/943/1815

Goldberg, A. E. (2018). *Transgender students in higher education.* The Williams Institute, UCLA School of Law. https://escholarship.org/uc/item/4p22m3kx

Goldberg, A. E., Beemyn, G., & Smith, J. Z. (2019). What is needed, what is valued: Trans students' perspectives on trans-inclusive policies and practices in higher education. *Journal of Educational and Psychological Consultation, 29*(1), 27–67. https://doi.org/10.1080/10474412.2018.1480376

Goldberg, A. E., & Kuvalanka, K. (2019). Transgender graduate students' experiences in higher education: A mixed-methods exploratory study. *Journal of Diversity in Higher Education, 12*(1), 38–51. https://psycnet.apa.org/doi/10.1037/dhe0000074

Goldberg, A. E., Smith, J. Z., & Beemyn, G. (2020). Trans activism and advocacy among transgender students in higher education: A mixed methods study. *Journal of Diversity in Higher Education, 13*(1), 66–84. https://psycnet.apa.org/doi/10.1037/dhe0000125

Gonzales, G., de Mola, E. L., Gavulic, K. A., McKay, T., & Purcell, C. (2020). Mental health needs among lesbian, gay, bisexual, and transgender college students during the COVID-19 pandemic. *Journal of Adolescent Health, 67*(5), 645–648. https://doi.org/10.1016/j.jadohealth.2020.08.006

Hendricks, M. L., & Testa, R. J. (2012). A conceptual framework for clinical work with transgender and gender nonconforming clients: An adaptation of the Minority Stress Model. *Professional Psychology: Research and Practice, 43*(5), 460–467. https://psycnet.apa.org/doi/10.1037/a0029597

James, C. E. (2021). Racial inequity, COVID-19 and the education of Black and other marginalized students. In *Impacts of COVID-19 in Racialized Communities* (pp. 36-44). Royal Society of Canada.

Kimble-Hill, A. C., Rivera-Figueroa, A., Chan, B. C., Lawal, W. A., Gonzalez, S., Adams, M. R., Heard, G. L., Gazley, J. L., & Fiore-Walker, B. (2020). Insights gained into marginalized students access challenges during the COVID-19 academic response. *Journal of Chemical Education, 97*(9), 3391–3395. https://doi.org/10.1021/acs.jchemed.0c00774

Kravitz, M. (2020, July 14). The gender binary is a tool of white supremacy. *An Injustice! Magazine.* https://aninjusticemag.com/the-gender-binary-is-a-tool-of-white-supremacy-db89d0bc9044

Lawrence, M., & Mckendry, S. (2019). *Supporting transgender and non-binary students and staff in further and higher education: Practical advice for colleges and universities.* Jessica Kingsley Publishers.

Lugones, M. (2016). The coloniality of gender. In *The Palgrave handbook of gender and development* (pp. 13-33). Palgrave Macmillan, London. http://doi.org/10.1007/978-1-137-38273-3_2

Maughan, L., Natalier, K., & Mulholland, M. (2022). Institutional transphobia: barriers to transgender research in early years education. *Gender and Education*, 721–737. https://doi.org/10.1080/09540253.2022.2057930

Mavhandu-Mudzusi, A. H., Mudau, T. S., Shandu, T., & Dorah, N. N. (2021). Transgender student experiences of online education during COVID-19 pandemic era in rural Eastern Cape area of South Africa: A descriptive phenomenological study. *Research in Social Sciences and Technology*, 6(2), 110–128.

McLemore, K. A. (2018). A minority stress perspective on transgender individuals' experiences with misgendering. *Stigma and Health*, 3(1), 53–64. https://psycnet.apa.org/doi/10.1037/sah0000070

McPhail, D., Rountree-James, M., & Whetter, I. (2016). Addressing gaps in physician knowledge regarding transgender health and healthcare through medical education. *Canadian Medical Education Journal*, 7(2), e70. http://dx.doi.org/10.36834/cmej.36785

Meyer, I. H. (2003). Prejudice, social stress, and mental health in lesbian, gay, and bisexual populations: Conceptual issues and research evidence. *Psychological Bulletin*, 129(5), 674–697. https://doi.org/10.1037%2F0033-2909.129.5.674

Omowale, J. (2021, August 18). *Colonialism still affects how black and indigenous people see gender.* Them. https://www.them.us/story/colonialism-black-and-indigenous-people-gender-identity

Phirangee, K., & Malec, A. (2017). Othering in online learning: An examination of social presence, identity, and sense of community. *Distance Education*, 38(2), 160–172. https://doi.org/10.1080/01587919.2017.1322457

Pryor, J. T. (2015). Out in the classroom: Transgender student experiences at a large public university. *Journal of College Student Development*, 56(5), 440–455. https://doi.org/10.1353/csd.2015.0044

Rankin, S. R. (2006). LGBTQA students on campus: Is higher education making the grade? *Journal of Gay & Lesbian Issues in Education, 3*(2-3), 111–117. http://dx.doi.org/10.1300/J367v03n02_11

Schneider, F. (2010). Where do we belong? Addressing the needs of transgender students in higher education. *The Vermont Connection, 31*(1), 11. https://scholarworks.uvm.edu/tvc/vol31/iss1/11/

Seelman, K. L. (2014). Recommendations of transgender students, staff, and faculty in the USA for improving college campuses. *Gender and Education, 26*(6), 618–635. https://scholarworks.gsu.edu/ssw_facpub/58

Selander, J., & Tidball, S. (2020). Creating and supporting an inclusive student services experience for trans and non-binary students. *College and University, 95*(2), 49–52.

Siegel, D. P. (2019). Transgender experiences and transphobia in higher education. *Sociology Compass, 13*(10), e12734. https://doi.org/10.1111/soc4.12734

Sinclair-Palm, J., & Chokly, K. (2022). 'It's a giant faux pas': Exploring young trans people's beliefs about deadnaming and the term deadname. *Journal of LGBT Youth*, 1–20.

Swanbrow Becker, M. A., Nemeth Roberts, S. F., Ritts, S. M., Branagan, W. T., Warner, A. R., & Clark, S. L. (2017). Supporting transgender college students: Implications for clinical intervention and campus prevention. *Journal of College Student Psychotherapy, 31*(2), 155–176. https://doi.org/10.1080/87568225.2016.1253441

Whitley, C. T., Nordmarken, S., Kolysh, S., & Goldstein-Kral, J. (2022). I've been misgendered so many times: Comparing the experiences of chronic misgendering among transgender graduate students in the social and natural sciences. *Sociological Inquiry, 31*(2), 155–176. https://doi.org/10.1111/soin.12482

APPENDIX A: Survey Questions

1. Please check this box to indicate that you are trans, gender-diverse, gender non-conforming, and/or non-binary. If this does not apply to you, please do not complete the survey.
    [ ]

2. Which year of study are you currently in at McMaster University?
    a. Year 1
    b. Year 2
    c. Year 3
    d. Year 4
    e. Year 5+

f. Master's
   g. PhD
   h. MD
   i. Other
   j. Prefer not to say
3. Which faculty are you currently a student in?
   a. Humanities
   b. Social Sciences
   c. Science
   d. Health Sciences
   e. Engineering
   f. DeGroote School of Business
   g. Other
   h. Prefer not to say
4. Please indicate your program of study. If you prefer not to say, please skip this question.
5. As a transgender, non-binary, or gender-diverse student at McMaster, relative to your experience with in-person classes, how would you describe your experience with online classes?
   a. Very Positive
   b. Positive
   c. Neutral
   d. Negative
   e. Very Negative
6. Based on your answer above, please elaborate on why your experience has been especially positive or negative if you feel comfortable.
7. How frequently have you felt uncomfortable or unsafe while navigating online learning technologies (e.g. participating in recorded lectures/seminars, turning your camera/audio on during synchronous online classes, etc.)?
   a. Almost Always or Always
   b. Often or Very Often
   c. Occasionally
   d. Rarely
   e. Almost Never or Never
8. How frequently have you been deadnamed or misgendered while navigating online learning technologies/platforms?
   a. Often or Very Often
   b. Occasionally
   c. Rarely
   d. Never
9. Of the online platforms that you have engaged with as a student at McMaster, which of these would you describe as being accessible and safe for you as a transgender, non-binary, or gender-diverse student?

a. Microsoft Teams: Very Unsafe/Inaccessible, Relatively Unsafe/Inaccessible, Neutral, Relatively Safe/Accessible, Very Safe/Accessible, N/A (do not use)
b. Zoom: Very Unsafe/Inaccessible, Relatively Unsafe/Inaccessible, Neutral, Relatively Safe/Accessible, Very Safe/Accessible, N/A (do not use)
c. Webex: Very Unsafe/Inaccessible, Relatively Unsafe/Inaccessible, Neutral, Relatively Safe/Accessible, Very Safe/Accessible, N/A (do not use)
d. Google Meets: Very Unsafe/Inaccessible, Relatively Unsafe/Inaccessible, Neutral, Relatively Safe/Accessible, Very Safe/Accessible, N/A (do not use)
e. Discord: Very Unsafe/Inaccessible, Relatively Unsafe/Inaccessible, Neutral, Relatively Safe/Accessible, Very Safe/Accessible, N/A (do not use)
f. Skype: Very Unsafe/Inaccessible, Relatively Unsafe/Inaccessible, Neutral, Relatively Safe/Accessible, Very Safe/Accessible, N/A (do not use)
g. Other (Insert Name): Very Unsafe/Inaccessible, Relatively Unsafe/Inaccessible, Neutral, Relatively Safe/Accessible, Very Safe/Accessible, N/A (do not use)
h. None of the Above
10. Follow-up: If you are comfortable sharing, please elaborate on why the platform(s) chosen in the previous question are preferred.
11. Have you been able to bring any concerns about online learning technologies to instructors, faculty members, teaching assistants, staff, or other administrators at the university?
    a. Yes
    b. No
12. If yes, what did this conversation look like? If no, what, if anything, has acted as a barrier to having this conversation?
13. What are some benefits of the online school setting as a transgender, non-binary, or gender-diverse student? Is there anything specific in the online setting that has made you realize what is lacking for transgender students during regularly scheduled in-person classes at McMaster?
14. What are some disadvantages of online learning that are particularly significant or salient as a trans, non-binary, or gender-diverse student? Is there anything specific in the online setting that has been of particular concern to you?
15. What are some ways in which instructors, teaching assistants, and staff can build more trans-inclusive spaces in their online classes? In other words, what would you want instructors, TAs, and staff to know while preparing their online classes?

APPENDIX B: Follow-Up Interview Questions

1. As a transgender, non-binary, or gender-diverse student at McMaster, relative to your experience with in-person classes, how would you describe your experience with online learning?

2. What are some particular concerns that have arisen for you during the transition to online classes and the use of online learning technologies?

3. How has the use of online learning platforms impacted your comfortability in participating/engaging in your classes, if at all? What has this looked/felt like?

4. Have you been able to discuss these concerns with anyone? If so, what did this look like?

5. Are there any positives that have come out of online learning? What, if anything, could be taken from online learning and applied to the transition into blended or in-person learning?

6. What are some ways in which instructors can build more trans-inclusive spaces in their online classes?

What Privacy? Online Privacy Culture and the Role of Libraries in Digital Information Literacy

Hannah Lee

The COVID-19 pandemic revealed how modern society requires online connectivity to function, and it also revealed many cracks in internet data sharing and privacy issues. When internet society first developed in the early 2000s, few could have predicted how much data and privacy individuals would relinquish for the sake of easy access to information and products. In modern times, technology companies make sharing and connecting through the different online apps and platforms simple, and it is hard to imagine using an app without an option to log in through Apple ID, Google, or Facebook. Prominent tech companies like Facebook, Twitter, Apple, and Amazon so often misuse and abuse their users' data that privacy breaches are almost the norm rather than the exception. Many of the services these companies provide do not cost money to use; however, the tradeoff for utilizing social media platforms and tech companies' free services is the willingness to have one's user data sold. But what data are individuals providing to tech companies? Is the user's data legally protected? In the United States, legislation meant to protect privacy stems from the decades-old Digital Millennium Copyright Act (1998) or, more recently, the California Consumer Privacy Act (CCPA) of 2018. However, legislation can do little if people do not know their digital information and privacy rights, which are often buried under complex legalese in end-user license agreements.

Librarians in higher educational institutions are in a unique position of understanding digital privacy issues by working with individuals' data, configuring library services, and discussing library and information science scholarship. Historically, libraries have valued patron privacy as a foundation of intellectual freedom (American Library Association, 2002) and strive to protect this privacy. While many libraries have been forced to comply with searches requiring libraries to disclose patron information (e.g., warrants, Patriot Act), some libraries have also warned patrons about such possibilities (Matz, 2008; Starr, 2004). Academic libraries have additional laws to comply with, compared to public libraries, where privacy laws in higher education impact students via FERPA. FERPA (Family Educational Rights and Privacy Act, 1974) guarantees that a student's educational privacy transfers from parent to student at 18 years old. Even though institutions may follow FERPA to the letter of the law and might try to comply and protect students' privacy data, the actual practice of FERPA is not always possible in a digital economy (Abbott, 2022; Brown & Klein, 2020; Inouye & Agnello, 2015; Lowenstein, 2016; Schrameyer et al., 2016).

Patron privacy in libraries has changed over the decades, modeling technological advances in libraries. As libraries shifted from being paper-based to technology-based, libraries have become more reliant on digital tools and services to meet community demands, with one prominent example being the development of open public access catalogs (OPACs). Today's complex discovery tools require multiple vendors and integrations, combining library services with a digital environment that includes restrictions like copyright, technical limitations, and data privacies.

The value of patron information is not limited to the sphere of higher education. While there is an emphasis in higher education on the value of information that comes from publication or research, individuals produce valuable information (Association of College & Research Libraries, 2015) as members of society. Being an informed citizen of online society means being aware of the value of our private information, what rights we give away, why we give our privacy away, and the consequences of our actions. The focus of this chapter is to reflect on the invasive state of current data privacy practices and how librarians in higher education can be an information source for privacy rights.

Background

The dictionary definition of privacy is simple in that there is an expectation of "being alone, undisturbed, or free" from attention (Oxford University Press, n.d.). Scholars debate the nature of privacy and the agency that individuals have in controlling information about themselves or freedom from external intrusion (ALA Office for Intellectual Freedom, 2010; Bélanger & Crossler, 2011; Campbell & Cowan, 2016; Kenyon & Richardson, 2006; Rotenberg et al., 2015; Sloot & Groot, 2018). The implications of individual privacy have evolved alongside technology and documentation methods.

Maintaining one's privacy was much easier before the computer age. Unless explicitly spoken, written, or observed in public, individuals did not have to worry too much about exposing their private information to the world. Without efforts to maintain memories through writing, visual media, or oral traditions, memories were short-lived. Even with early cameras and recording devices, there was a barrier of entry to have access to these tools. Either processing film took a long time, or the equipment was too expensive for the average person.

Libraries and Privacy

During this pre-computer era, libraries were limited by the ability to keep physical records. Using check-out cards and due date slips, books and other media checked out from the library could be traced back to its user through a patron's handwritten name or ID (Surace, 1970). However, patron privacy could easily be preserved by writing over names on the checkout cards or eliminating paperwork linking patron IDs to the material. This era of library circulation would also require a deeper search to connect patrons to their checked-out material; someone would need to find the original book to link it to an individual. If there were multiple copies of the same item, then there would be the additional complication to link to the same copy that a patron checked out. However, privacy in libraries began to change with the introduction of public computers.

One of the places people were able to gain access to computers in the early days of the internet was the library. Starting with early computers intended to help with basic functions like payroll and accounting, there was growing recognition for more complicated tasks using computers (Allen, 2014; Arms, 2012). In the 1960s, the Library of Congress investigated the possibility of a machine-based form of information storage and retrieval, eventually giving the library sciences what is known today as Machine Readable Cataloging, or MARC (Avram, 1975). While libraries had library classification systems like the Dewey Decimal System (in many public libraries) or the Library of Congress Call Number System (in many academic libraries), it was difficult to search for items unless you knew the subject-based term needed to begin the search. Even card catalogs did not have a standard means of organizing; while sorting by the author's last name was common, it was not the only method of card catalog organization (Pachefsky, 1969). These physical organization discrepancies would lead to the development of computer-based catalogs to aid in searching.

Soon after the Library of Congress considered (and approved) the idea of using MARC came the Online Public Access Catalog (OPAC). Although the term "OPAC" came into use in the 1980s, the basic idea was to have a public-facing (i.e., patron-accessible) catalog of a library's holdings (Wells, 2020). Most of the early OPACs concentrated on providing local records for local patrons; in other words, you could only search for things using the library's OPAC in the library where you were physically located (or call to ask).

The OPAC made searching via keywords easier, but the growth of the internet and the availability of different types of information required complex systems to address an increasingly complex digital information society. This complexity brought on the need for what libraries call "discovery systems." Discovery systems allowed patrons to go beyond a local library's catalog record to enable basic functions present in internet searching (e.g., spelling corrections, autofill suggestions) while connected to previously separate systems like databases. Instead of searching specific databases for scholarly articles (requiring some level of working knowledge of each discipline or databases in general) and a separate system for the library catalog, discovery systems allowed for multiple functions from a single platform (Dahl, 2009; Giza, 2022). The interconnectedness of discovery systems often happens through single sign-on (SSO) credentials and proxies tied to an individual's email or school username. However, if systems do not have built-in mechanisms anonymizing users' data, then privacy breaches can occur. Libraries can link individual users with their network access, especially as databases are trying to curtail piracy and track users through proxy networks. While discovery systems make accessing information convenient through a single portal, the library and its librarians must ensure that patron privacy remains intact.

The library profession advocates for patron privacy as a part of its Bill of Rights (American Library Association, 2002), but digital technologies add challenging layers (Gardner, 2002; Hess et al., 2015). With the increase in technical functionality of libraries came an accompanying increase in digital privacy innovations and accompanying concerns. Rather than using physical checkout cards, books and other materials now have barcodes linked to electronic records. With digital recordkeeping, libraries take care to only keep records of what patrons check out *while* the items are checked out; typically, the default for OPAC and discovery systems is to delete (or destroy) a patron's checkout history. While there are metrics in place to show how many times a single item may have circulated, library systems do not connect them to the patron (Klinefelter, 2007; Pekala, 2017).

Even with access to electronic resources like eBooks, journal articles, databases, and other electronic materials paid for by the library, these systems often tie a username or library card number with a limit to how long that record exists. Typically, through the use of proxy networks capable of authenticating access to electronic resources, a patron can get access without giving too much personal information. Of course, if the patron's ID username is easily identifiable (e.g., firstname.lastname), identifying individuals would be easy if not for the technical limits of how long libraries keep patron information (Murray, 2001; Shabtai et al., 2013; Wang et al., 2016).

As interoperability and ease of access to various platforms grow, the library is one of the few places that does not tie usage to a social media platform.

Libraries have their own social media accounts for marketing purposes, but a patron would not be able to "Login using Facebook" or "Use your Google Account" as options for logging in to library services. While the minimum for obtaining a library card is typically an email address (more for notifications and as a means of contacting), libraries do not connect with a patron's social media to build a profile in the same way a technology company develops large datasets.

This separation of social media and library accounts means individuals can keep their library habits private without worrying that libraries will sell their information or similarly disclose their information. In an internet era requiring so much disclosure of personal information, dedication to privacy means that libraries are a place where individuals can seek knowledge with assured privacy protections (Cooke, 2018; Rubel, 2014). Privacy in libraries becomes a critical component of personal freedoms, particularly for marginalized people or people in communities actively censoring information (Spilka, 2022).

That is not to say that privacy invasions do not occur. A study of public libraries and library vendors found that many vendors did not meet the professional standards for using and handling users' information (Lambert, et al., 2015). Additionally, a 2010 study also revealed the same concerns about vendors, adding that although library vendors were transparent about their practices, little could be done by library patrons or libraries (Magi, 2010). However, more recent studies show that library vendors are catching up to the privacy needs of libraries and their users (McKinnon & Turp, 2022; Yoose, 2017). Although additional studies should be done in the future to ensure that users' privacies remain intact, there is a growing trend towards system-level privacies for individuals.

Technology and Privacy

Tech companies like Apple, Facebook, Google, and others create large data from their users. A Cisco report forecasted an increase in IP traffic from approximately 37,075 GB per second in 2016 to 107,291 GB per second in 2021 (2016). In contrast, an updated estimate from Cisco and the World Bank Group estimates 150,000 GB per second of data by 2022 (2019; 2021). Of course, the initial forecasts could not have predicted the COVID-19 pandemic, which prompted an exponential increase in internet traffic. Along with data showing that 80% of adults in the United States use some form of social media several times throughout the day, it is easy to see how much data individuals generate daily (Auxier & Anderson, 2021).

The amount of data that individuals generate is important for tech companies. Companies use this data to run and sell advertisements on their sites. Because social media is free to use, companies make most of their profits from advertising revenue (Leetaru, 2018). Mark Zuckerberg, founder and CEO of Facebook, responded to Senator Orrin Hatch's question of how tech companies make money with, "Senator, we run ads" (CSPAN, 2018). Senator Hatch's lack of understanding of how companies sell users' data is indicative of how little the public knows (past and present) about how major tech companies operate. After all, these tech companies are part of a billion-dollar industry that commodifies user data in exchange for services. But how did we get to this point?

In the relatively early days of internet society, there were few protections in place that would ensure a user's privacy. While European countries enabled data protection statutes before 2000 (Sian, 2012; Solove, 2006), the first and most lasting piece of legislation in the United States is the Digital Millennium Copyright Act (DMCA) of 1998. While the DMCA protects internet service providers from copyright infringement, individual internet subscribers could still be tied to requests for information like copyright infringement (Katyal, 2004; Penney, 2019). Even with the ability to connect IP addresses to people, there are errors. As one family in Kansas found out, 600 million IP addresses were associated with their rented farm address (Farivar, 2016), prompting repeated queries from law enforcement and other officials.

Perhaps one of the largest discoveries of online privacy infringement happened in 2013 when former NSA analyst Edward Snowden revealed the government-run PRISM program that allowed unfettered government surveillance of corporations and private citizens (Farrell & Newman, 2019; Lucas, 2014; Macnish, 2018). PRISM was a direct result of the 2001 USA PATRIOT Act and the 2007 Protect America Act, laws that allowed for a massive overreach of how much the government could use digital tools to invade the daily lives of people.

The Patriot Act and the Protect America Act affect users through the government's interpretation of data and its relation to the US border. Data exists on servers, which are computer hardware containing the data that make up the ones and zeros of the internet. Additionally, servers are often located outside of the United States where they may be cheaper to build and where the energy required to cool and maintain the equipment is more cost-effective. If you download a photo you find on the internet and share it with friends, the data for that photo could exist anywhere from California to Maine, or in countries in Europe, South America, or Asia. The way the Patriot Act and the Protect America Act have been interpreted is that any digital data crossing the US border is subject to surveillance. This specific example of privacy intrusion by the government has been done in the name of national security. In the late 2000s/early 2010s, the United States experienced major technological events like Edward Snowden's data leak. Consequently, Congress passed the Personal Data Privacy and Security Act of 2009, increasing punishment for identity theft and other such privacy and security breaches. Otherwise, data privacy rights were largely under the purview of individual states.

Not until the passage of the General Data Protection Regulation (GDPR) in 2016 and the California Consumer Privacy Act (CCPA) of 2018 would there be major steps toward online privacy. The GDPR is a European Union law regulating data protection and privacy; the CCPA gives individuals more control over their data and how companies can use that data. With so many tech companies located in California and a significant portion of consumers/customers in Europe, these statutes have helped establish de facto standards of online privacy and data privacy (Barrett, 2019; *California Consumer Privacy Act (CCPA)*, 2018; Fazlioglu, 2020; Rakoski, 2021; Regulation 2016/679).

Another event that necessitated greater privacy regulations occurred in 2017 with the Cambridge Analytica (CA)-Facebook incident. This incident demonstrated the need for the increased privacy regulations that the CCPA provided. Cambridge Analytica (CA) was a company that provided resources and services for political campaigns around the globe. However, Facebook gave CA unrestricted access to users' data and other personally identifiable information (Isaak & Hanna, 2018; Shipman & Marshall, 2020). The ability to micro-target ads and information to individuals only became possible with the data that users unknowingly gave to private companies. What made the incident so egregious was that Cambridge Analytica used Facebook users' information without letting them know *the purpose* of their information collection, and CA has since been tied to political interference in the 2014 US Midterm elections, the 2016 Brexit referendum, and the 2016 US Presidential election (Hinds et al., 2020; Richterich, 2018; ur Rehman, 2019). After all, it is one thing to agree to limited information in exchange for free services; it is another to allow companies enough individual information to alter entire elections.

Data brokers add another layer of data privacy intrusions to existing privacy concerns. While the idea of large data sets based on metadata or non-identifiable information seems safe to use, the practical reality is that individuals are identifiable. Data brokers and others in the practice of trading people's information can use the data that they buy, sell, and gather through targeted ads (e.g., Facebook) for election misinformation (Otto et al., 2007; Rostow, 2017). Moreover, the longevity of digital data prevents the right to be forgotten in many cases in the United States. While Europe and the European Union may allow individuals to ask Google and other online platforms to "forget" (i.e., remove) certain information about themselves, the United States does not enjoy those same privacy protections (Gajda, 2018; Rosen, 2011; Tsesis, 2014). Additionally, "public figures" exist under the "right to know" provisions of information (Shackelford, 2012; Yanisky-Ravid & Lahav, n.d.).

Discussion

So how do discussions on data privacy, libraries in higher education, and digital literacy begin? It's complicated. There needs to be legislation and general best practices capable of protecting individuals' data privacy, and there also needs to be accountability and real consequences for those who misuse and abuse individuals' rights to privacy. Ultimately, this is a complex issue without a single solution.

One way of addressing the complexities of data privacy is to start with individuals and provide them with the education to be information-literate citizens. Information literacy is not only an important part of higher education in developing critical thinking and problem-solving skills, but it also prepares students to be a part of an information-rich society. Never has there been more information circulated and generated on a regular basis, and data and information circulation will only continue to increase exponentially in the future (Eisenberg et al., 2004; Koltay, 2011; Rockman, 2004; Ross et al., 2016). *Digital* information literacy is adding a layer of digital data and information to the existing scaffold of information literacy in a complex digital world (Jeffrey et al., 2011; Sparks, et al., 2016).

Libraries have been the entity in higher education charged with much of the information literacy pedagogy. From the humanities to the sciences and everywhere in between, the university or academic library remains at the core of the institution when it comes to information literacy (Hicks & Lloyd, 2021; Sample, 2020; Sparks, et al., 2016). With libraries being systems that understand the value of privacy and caution when it comes to innovations in information technologies, librarians tend to understand the theoretical ramifications of data privacy along with its practical applications. Librarians' professional experience and educational qualifications make them ideal candidates for teaching information literacy.

Higher educational institutions require librarians to have a master's degree, either an MLIS (Master of Library and Information Science) or some equivalent. Additionally, to be accredited by the American Library Association (2008), a master's program needs to meet several core competencies in research, technology, information resources, and other related domains. In addition, there are many specializations a person can focus on both as a core function for their work and for scholarship. Data management librarians, information literacy librarians, and instruction librarians are but a few of the different titles held by librarians who regularly deal with information literacy.

Libraries are not in the business of selling data (private or otherwise). Instead, libraries focus on keeping user data private, and there has always been a professional emphasis on supporting individual privacy. This exists within scholarly pedagogies emphasizing good data practices and the de-identification of individuals. While additional revenue streams are always welcome in the library given the rising cost of journal and database access, selling student data would be against traditions and practices in higher education libraries.

These libraries and librarians are a potential source for teaching digital information literacy, not only because of their establishment as an information and knowledge center at the university but also because they are already providing these types of literacy services and teaching. There are already many opportunities and resources for collaboration between librarians and the university. The Johns Hopkins Digital Literacy Resources guide, the University of British Columbia's Digital Tattoo project, Syracuse University's Center for Digital Literacy, and the University of Pennsylvania's Digital Literacy Fellows program are just some of the examples where libraries/librarians lead or collaborate to bring digital literacy to higher education. Additional digital literacy services performed by librarians include regular instruction and reference sessions that connect students and faculty with the library (Kocevar-Weidinger et al., 2019; Martzoukou & Sayyad Abdi, 2017; Meyers, et al., 2013). All these practical efforts are on top of the scholarship that librarians produce that look at historical, current, and future practices of digital literacies (Withorn et al., 2021).

What the average user needs to realize is that there are processes in place to protect their data. Even though it can be annoying to have to review privacy settings for websites or perform regular audits of what platforms have access to your email and Facebook data, a well-informed, data information literate person will know *why* these measures are in place and appreciate the effort it took to get here. Until there is a critical mass of people and companies who build information technologies and use them with data privacy in mind, education is another way to address the issue. One way to start that journey is to begin at the library with its librarians.

References

ALA Office for Intellectual Freedom. (2010). *Privacy and freedom of information in 21st-century libraries*. American Library Association.

Abbott, H. (2022). How data stewards make decisions to protect or disclose student information: Toward consistent criteria. *College & University*, *97*(1), 2–9.

Allen, E. (2014, January 15). *A half century of library computing*. Library of Congress Blog. https://blogs.loc.gov/loc/2014/01/a-half-century-of-library-computing/

American Library Association. (2002). *Privacy: An interpretation of the library bill of rights* https://www.ala.org/advocacy/intfreedom/librarybill/interpretations/privacy

American Library Association. (2008, June 10). *Core competences* https://www.ala.org/educationcareers/careers/corecomp/corecompetences

Arms, W. Y. (2012). The 1990s: The formative years of digital libraries. *Library Hi Tech*, *30*(4), 579–591. https://doi.org/10.1108/07378831211285068

Association of College & Research Libraries. (2015, February 9). *Framework for information literacy for higher education.* https://www.ala.org/acrl/standards/ilframework

Auxier, B., & Anderson, M. (2021). *Social media use in 2021*. Pew Research Institute. https://www.pewresearch.org/internet/2021/04/07/social-media-use-in-2021/

Avram, H. D. (1975). *MARC; its history and implications*. Superintendent of Documents, U.S. Government Printing Office. http://eric.ed.gov/ERICWebPortal/detail?accno=ED127954

Barrett, C. (2019). Are the EU GDPR and the California CCPA becoming the de facto global standards for data privacy and protection? *The SciTech Lawyer*, *15*(3), 24–29.

Bélanger, F., & Crossler, R. E. (2011). Privacy in the digital age: A review of information privacy research in information systems. *MIS Quarterly*, *35*(4), 1017–1041. https://doi.org/10.2307/41409971

Brown, M., & Klein, C. (2020). Whose data? Which rights? Whose power? A policy discourse analysis of student privacy policy documents. *Journal of Higher Education*, *91*(7), 1149–1178. https://doi.org/10.1080/00221546.2020.1770045

California Consumer Privacy Act (CCPA). (2018, October 15). State of California Department of Justice, Office of the Attorney General. https://oag.ca.gov/privacy/ccpa

Campbell, D. G., & Cowan, S. R. (2016). The paradox of privacy: Revisiting a core library value in an age of big data and linked data. *Library Trends*, *64*(3), 429–511. http://dx.doi.org/10.1353/lib.2016.0006

Cisco. (2016). *Global –2021 forecast highlights*. Cisco. https://www.cisco.com/c/dam/m/en_us/solutions/service-provider/vni-forecast-highlights/pdf/Global_2021_Forecast_Highlights.pdf

Cisco. (2019). *Cisco visual networking index: Forecast and trends, 2017–2022*. Cisco. https://twiki.cern.ch/twiki/pub/HEPIX/TechwatchNetwork/HtwNetworkDocuments/white-paper-c11-741490.pdf

Cooke, L. (2018). Privacy, libraries and the era of big data. *IFLA Journal, 44*(3), 167–169. https://doi.org/10.1177/0340035218789601

CSPAN. (2018, April 10). *User Clip: Conversation between Mark Zuckeberg and Senator Orring Hatch—"Senator, we run ads"* [Video]. C-SPAN. https://www.c-span.org/video/?c4726758/user-clip-conversation-mark-zuckeberg-senator-orring-hatch-senator-run-ads

Dahl, M. (2009). The evolution of library discovery systems in the web environment. *OLA Quarterly, 15*(1), 5–9. https://doi.org/10.7710/1093-7374.1229

Digital Millennium Copyright Act, H.R. 2281, 105th Cong. (1998). https://www.congress.gov/bill/105th-congress/house-bill/2281

Eisenberg, M. B., Lowe, C. A., & Spitzer, K. L. (2004). *Information literacy: Essential skills for the information age*. ERIC.

Farivar, C. (2016, August 10). Kansas couple sues IP mapping firm for turning their life into a "digital hell." *Ars Technica*. https://arstechnica.com/tech-policy/2016/08/kansas-couple-sues-ip-mapping-firm-for-turning-their-life-into-a-digital-hell/

Farrell, H., & Newman, A. L. (2019). Of privacy and power: The transatlantic struggle over freedom and security. In *Of Privacy and Power*. Princeton University Press.

Fazlioglu, M. (2020). The United States and the EU's General Data Protection Regulation. In *Data Protection Around the World* (pp. 231–248). T.M.C. Asser Press. https://doi.org/10.1007/978-94-6265-407-5_10

Gajda, A. (2018). Privacy, press, and the right to be forgotten in the United States. *Washington Law Review, 93*(1), 201-264. https://ssrn.com/abstract=3144077

Gardner, C. (2002). Fact or fiction: Privacy in American libraries. *Proceedings of the 12th Annual Conference on Computers, Freedom and Privacy*, 1–5. https://doi.org/10.1145/543482.543503

Giza, P. (2022). Automated discovery systems, part 1: Historical origins, main research programs, and methodological foundations. *Philosophy Compass, 17*(1). https://doi.org/10.1111/phc3.12800

Hess, A. N., LaPorte-Fiori, R., & Engwall, K. (2015). Preserving patron privacy in the 21st century academic library. *The Journal of Academic Librarianship, 41*(1), 105–114. https://doi.org/10.1016/j.acalib.2014.10.010

Hicks, A., & Lloyd, A. (2021). Deconstructing information literacy discourse: Peeling back the layers in higher education. *Journal of Librarianship and Information Science, 53*(4), 559–571. https://doi.org/10.1177/0961000620966027

Hinds, J., Williams, E. J., & Joinson, A. N. (2020). "It wouldn't happen to me": Privacy concerns and perspectives following the Cambridge Analytica scandal. *International Journal of Human-Computer Studies, 143*, 102498. https://doi.org/10.1016/j.ijhcs.2020.102498

Inouye, T. M., & Vincent Agnello, J. D. (2015). Higher education industry consolidation: Where does it leave students? *Journal of Religion and Business Ethics, 4*(1), 1–14.

Isaak, J., & Hanna, M. J. (2018). User data privacy: Facebook, Cambridge Analytica, and privacy protection. *Computer, 51*(8), 56–59. https://doi.org/10.1109/MC.2018.3191268

Jeffrey, L., Hegarty, B., Kelly, O., Penman, M., Coburn, D., & McDonald, J. (2011). Developing digital information literacy in higher education: Obstacles and supports. *Journal of Information Technology Education: Research, 10*(1), 383–413. http://dx.doi.org/10.28945/1532

Katyal, S. K. (2004). Privacy vs. Piracy. *Yale Journal of Law & Technology, 7*.

Kenyon, A. T., & Richardson, M. (2006). *New dimensions in privacy law: International and comparative perspectives*. University Press.

Klinefelter, A. (2007). Privacy and library public services: Or, I know what you read last summer. *Legal Reference Services Quarterly, 26*(1-2), 253–279. https://doi.org/10.1300/J113v26n01_13

Kocevar-Weidinger, E., Cox, E., Lenker, M., Pashkova-Balkenhol, T., & Kinman, V. (2019). On their own terms: First-year student interviews about everyday life research can help librarians flip the deficit script. *Reference Services Review, 47*(2), 169–192. https://doi.org/10.1108/RSR-02-2019-0007

Koltay, T. (2011). The media and the literacies: Media literacy, information literacy, digital literacy. *Media, Culture & Society, 33*(2), 211–221. https://doi.org/10.1177/0163443710393382

Leetaru, Kalev. (2018, December 15). *What does it mean for social media platforms to "sell" our data?* Forbes. https://www.forbes.com/sites/kalevleetaru/2018/12/15/what-does-it-mean-for-social-media-platforms-to-sell-our-data/?sh=54620a082d6c

Lambert, A. D., Parker, M., & Bashir, M. (2015). Library patron privacy in jeopardy an analysis of the privacy policies of digital content vendors.

Proceedings of the *Association for Information Science and Technology, 52*(1), 1–9. https://doi.org/10.1002/pra2.2015.145052010044

Lowenstein, H. (2016). The great wall of FERPA: Surmounting a law's barrier to assurance of learning. *The Journal of Legal Studies Education, 33*(1), 129–164. https://doi.org/10.1111/jlse.12037

Lucas, G. R. (2014). NSA Management Directive #424: Secrecy and privacy in the aftermath of Edward Snowden. *Ethics & International Affairs, 28*(1), 29–38. https://doi.org/10.1017/S0892679413000488

Macnish, K. (2018). Government surveillance and why defining privacy matters in a post-Snowden world. *Journal of Applied Philosophy, 35*(2), 417–432. https://doi.org/10.1111/japp.12219

Magi, T. J. (2010). A content analysis of library vendor privacy policies: Do they meet our standards? *College & Research Libraries, 71*(3), 254–272.

Martzoukou, K., & Sayyad Abdi, E. (2017). Towards an everyday life information literacy mind-set: A review of literature. *Journal of Documentation, 73*(4), 634–665. https://doi.org/10.1108/JD-07-2016-0094

Matz, C. (2008). Libraries and the USA PATRIOT Act: Values in conflict. *Journal of Library Administration, 47*(3–4), 69–87. https://doi.org/10.1080/01930820802186399

McKinnon, D., & Turp, C. (2022). Are library vendors doing enough to protect users? A content analysis of major ILS privacy policies. *The Journal of Academic Librarianship, 48*(2). https://doi.org/10.1016/j.acalib.2022.102505

Meyers, E. M., Erickson, I., & Small, R. V. (2013). Digital literacy and informal learning environments: An introduction. *Learning, Media and Technology, 38*(4), 355–367. https://doi.org/10.1080/17439884.2013.783597

Murray, P. E. (2001). Library Web proxy use survey results. *Information Technology & Libraries, 20*(4), 172.

Otto, P. N., Antón, A. I., & Baumer, D. L. (2007). The choicepoint dilemma: How data brokers should handle the privacy of personal information. *IEEE Security & Privacy, 5*(5), 15–23. https://doi.org/10.1109/MSP.2007.126

Oxford University Press. (n.d.). Privacy. In *OED Online*. Retrieved May 14, 2022, from http://www.oed.com/view/Entry/151596

Pachefsky, R. (1969). Survey of the card catalog in medical libraries*. *Bulletin of the Medical Library Association, 57*(1), 10–20.

Penney, J. W. (2019). Privacy and legal automation: The DMCA as a case study. *Stanford Technology Law Review, 22*, 412–486.

Pekala, S. (2017). Privacy and user experience in 21st century library discovery. *Information Technology and Libraries, 36*(2), 48–58. https://doi.org/10.6017/ital.v36i2.9817

Rakoski, R. L. (2021). Navigating global privacy regulations. *Benefits Magazine, 58*(3), 50–56.

Regulation 2016/679 *General Data Protection Regulation (GDPR)*. Retrieved May 15, 2022, from https://gdpr-info.eu/

Richterich, A. (2018). How data-driven research fueled the Cambridge Analytica controversy. *Partecipazione e Conflitto, 11*(2), 528–543. https://doi.org/10.1285/i20356609v11i2p528

Rockman, I. F. (2004). *Integrating information literacy into the higher education curriculum: Practical models for transformation.* Jossey-Bass.

Rosen, J. (2011). The right to be forgotten. *Stanford Law Review Online, 64*, 88–92.

Ross, M., Perkins, H., & Bodey, K. (2016). Academic motivation and information literacy self-efficacy: The importance of a simple desire to know. *Library & Information Science Research, 38*(1), 2–9. https://doi.org/10.1016/j.lisr.2016.01.002

Rostow, T. (2017). What happens when an acquaintance buys your data: A new privacy harm in the age of data brokers. *Yale Journal on Regulations, 34*(2), 667–707.

Rotenberg, M., Scott, J., & Horwitz, J. (2015). *Privacy in the modern age: The search for solutions.* The New Press.

Rubel, A. (2014). Libraries, electronic resources, and privacy: The case for positive intellectual freedom. *The Library Quarterly, 84*(2), 183–208. https://doi.org/10.1086/675331

Sample, A. (2020). Historical development of definitions of information literacy: A literature review of selected resources. *The Journal of Academic Librarianship, 46*(2), 102116. https://doi.org/10.1016/j.acalib.2020.102116

Schrameyer, A. R., Graves, T. M., Hua, D. M., & Brandt, N. C. (2016). Online student collaboration and FERPA considerations. *TechTrends, 60*(6), 540–548. https://doi.org/10.1007/s11528-016-0117-5

Scott, W. R. (2007). *Organizations and organizing: Rational, natural, and open system perspectives* (1st ed.). Pearson Prentice Hall.

Shabtai, A., Morad, I., Kolman, E., Eran, E., Vaystikh, A., Gruss, E., Rokach, L., & Elovici, Y. (2013). IP2User -- Identifying the username of an IP address in network-related events. 2013 *IEEE International Congress on Big Data*, 435–436. https://doi.org/10.1109/BigData.Congress.2013.73

Shackelford, S. J. (2012). Fragile merchandise: A comparative analysis of the privacy rights for public figures. *American Business Law Journal, 49*(1), 125–208. https://doi.org/10.1111/j.1744-1714.2011.01129.x

Shipman, F., & Marshall, C. (2020, April). Ownership, privacy, and control in the wake of Cambridge Analytica. *Proceedings of the 2020 CHI Conference on Human Factors in Computing Systems.* CHI Conference 2020, Honolulu HI USA. https://doi.org/10.1145/3313831.3376662

Sian, R. (2012). Origins and historical context of data protection law. In Eduardo Ustaran (Ed.), *European privacy: Law and practice for data protection professionals.* International Association of Privacy Professionals.

Sloot, B. van der, & Groot, A. de. (2018). *The handbook of privacy studies: An interdisciplinary introduction.* University Press. https://doi.org/10.1515/9789048540136

Solove, D. J. (2006). A brief history of information privacy law (SSRN Scholarly Paper No. 914271). *Social Science Research Network.* https://papers.ssrn.com/abstract=914271

Sparks, J. R., Katz, I. R., & Beile, P. M. (2016). Assessing digital information literacy in higher education: A review of existing frameworks and assessments with recommendations for next-generation assessment. *ETS Research Report Series, 2016*(2), 1–33. https://doi.org/10.1002/ets2.12118

Spilka, J. (2022). 377 Book challenges tracked by ALA in 2019--and the problem is growing: Book banning and its adverse effects on students. *Knowledge Quest, 50*(5), 30–33.

Starr, J. (2004). Libraries and national security: An historical review. *First Monday, 9*(12). https://doi.org/10.5210/fm.v9i12.1198

Surace, C. J. (1970). *Library circulation systems–an overview.* RAND Corporation.

Syracuse University. (n.d.). *Center for Digital Literacy.* https://www.digital-literacy.syr.edu/

Tsesis, A. (2014). The right to erasure: Privacy, data brokers, and the indefinite retention of data. *Wake Forest L. Rev., 49*, 433–484.

University of British Columbia. (n.d.). *Digital Tattoo.* https://digitaltattoo.ubc.ca/

University of Pennsylvania. (n.d.). *Hoesley Digital Literacy Fellows Program.* https://www.curf.upenn.edu/content/hoesley-digital-literacy-fellows-program

ur Rehman, I. (2019). Facebook-Cambridge Analytica data harvesting: What you need to know. *Library Philosophy and Practice.* 2497.

Wang, Y., Liu, T., Tan, Q., Shi, J., & Guo, L. (2016). Identifying users across different sites using usernames. *Procedia Computer Science, 80*, 376–385. https://doi.org/10.1016/j.procs.2016.05.336

Wells, D. (2020). *Online public access catalogues and library discovery systems* [Text]. https://www.isko.org/cyclo/opac

Welsh Medical Library. (n.d.). *Digital and Information Literacy.* Johns Hopkins University. https://browse.welch.jhmi.edu/teaching-learning-resources/digital-information-literacy

Withorn, T., Eslami, J., Lee, H., Clarke, M., Gardner, C. C., Springfield, C., Ospina, D., Andora, A., Castañeda, A., Mitchell, A., Kimmitt, J. M., Vermeer, W., & Haas, A. (2021). Library instruction and information literacy 2020. *Reference Services Review, 49*(3/4), 329–418. https://doi.org/10.1108/RSR-07-2021-0046

Yanisky-Ravid, S., & Lahav, B. Z. (n.d.). Public interest vs. private lives—Affording public figures privacy in the digital era: The three principle filtering model. *University of Pennsylvania Journal of Constitutional Law, 19*(5). https://ssrn.com/abstract=2931864

Yoose, B. (2017). Balancing privacy and strategic planning needs: A case study in de-identification of patron data. *Journal of Intellectual Freedom & Privacy, 2*(1), 15–22. https://doi.org/10.5860/jifp.v2i1.6250

Part 2: How Do We Take Care of Our Students, Ourselves?

Professional Identity and Digital Diligence

Angela Dixon and Amy Stalker

The transition to online teaching has dramatically increased educator visibility. Unfortunately, in some cases, that increased attention brings a disproportionate level of negative scrutiny. While adapting to significant changes in teaching modalities, educators must also consider the need to protect their privacy through the separation of their personal and professional identities. The ever-widening gap in socio-political opinion coupled with a worldwide pandemic has set the stage for hyper-scrutiny within personal and digital spaces. As an unexpected benefit of being abruptly forced online for daily communication, the public witnessed a dramatic improvement in collective technical skills. Upgrades to internet services and computer equipment served to further supplement that growth. However, this increase in online life also opened windows for broader exposure to subjects and personalities that some found at odds with their personal beliefs. The education system is one of many that has faced intense re-examination in this environment. Given the increase in exposure and the unpredictability of public response, educators should consider actively separating their personal identities from their professional identities while proactively assessing privacy risks. The terms *educator*, *teacher*, and *instructor* will be used interchangeably throughout this chapter. Though there are some notable differences between the K-12 environment and higher education, the focus of this chapter covers both arenas.

Preserving a positive professional identity is critical to a successful career as an educator. This manifests through what has been called a "persona." Major (2015) stated that a persona "represents a compromise between the role that a given individual is willing to play and the role that society expects" (p. 164). Maintaining that balance is tenuous at best. Unfortunately, these carefully crafted personas can be easily ruined. Be it maliciously or accidentally, a damaged reputation for an educator often negates the years of pre-career training and years of success inside the classroom. From the start of the American public education system, educators have been subjected to intense scrutiny by the public while simultaneously being held to higher moral standards. Through the early years of the twentieth century, teachers could be punished or dismissed altogether for getting married, having a child, or participating in any community-defined infractions, among other reasons (Pawlewicz, 2020). With increased internet usage during the lockdown period of the pandemic, educators began experiencing heightened attention to and judgment of their teaching methods and practices, including the expectation of unlimited availability (Weale, 2022). They are critiqued in a variety of public forums for what happens inside the classroom, such as their approach to course content, as well as those things that happen off school property outside of school hours, such as community activities and activism. The transition to online teaching during the pandemic continues to erode educators' expectations of privacy and agency to separate work life from personal on their terms.

Effects of the Pandemic

The COVID-19 pandemic instantly changed the way people worked and lived. The surrounding uncertainty regarding exposure to infection and unprecedented death pushed most aspects of everyday life online, including doctor's visits and funeral services (Kaffer, 2020). As the internet became the world's dominant means of communication, computer equipment sales skyrocketed as households sought to improve online connections and experiences. Sales of personal computers saw their most significant growth in a decade, with largely Chromebooks filling the almost-immediate demands of online learning (Armental, 2020). Quick pivots to working almost exclusively online drove laptop and desktop sales to exceed $302 million in 2020 (Pressman, 2021). The additions or upgrades to web cameras, camera lighting, microphones, speakers, and high-resolution monitors, coupled with a tremendous increase in technology dependency, led to a quick uptick in user skillset and confidence. Being online was no longer a preference but a necessity. Even users ages 50 years and older contributed to these marked increases by 8-15% in their own adoption of key technologies, such as smartphone ownership and social media use, compared to a decade ago (Faverio, 2022). This increase in users, particularly on social media, added to the number of voices expressing opinions and in turn, gave everyone a voice on issues from mask requirements to vaccinations to in-person vs. online learning (Associated Press, 2020).

Improved technical knowledge, increased time online, and the cultivation of new online personas quickly expanded the potential audience of educators outside the confines of the classroom roster and educational institutions. The audience now included the court of public opinion and excessive media coverage – where the rules seem to follow the old "shoot first, ask questions later" approach. For the first time and on a grand scale, the online classroom provided a viewing window for those not enrolled as students. Family and guardians were able to observe, attend and comment on learning happening in a digital arena rife with privacy concerns and access issues. This change emboldened some parents to "zoom bomb" active online sessions by breaking in to ask questions or to refute the teacher (Jargon, 2020). Inevitably, this new access point of instant feedback and concern led to conflicts such as online classroom management disagreements with parents, online bullying over the return to in-person learning, and full-scale targeting that could lead to professional burnout/demotion/job loss.

Challenges of the Online Classroom

The online classroom introduced a host of unforeseen privacy challenges for educators. Unlike a traditional face-to-face classroom, the online platform makes it impossible to reliably detect if someone not included in the class roster is lurking in the background. Lurking increased opportunities for misunderstanding and misinterpretation of class management, instructional modes, and even the content itself. Online learning sessions can be recorded and distributed without instructor knowledge even though subsequent edits might take the situation out of context. These possibilities almost demand that educators regularly inspect their recording environment before beginning each online session and consider what could be visually misconstrued in the background. What will students and parents be able to see in the background? Items include books, artwork, photos, flags, and anything else that observers could characterize as offensive or objectionable. Unfortunately, this can also include apparel like shirts, scarves, and hats that imply affiliation or support for various groups or movements. The potential of parents to misconstrue or incorrectly contextualize objects means there is value in depersonalizing everything when teaching online. For example, a high school teacher in Los Angeles was forced to flee her home due to death threats received after wearing a Black Lives Matter (BLM) t-shirt while teaching an online English session (Agrawal, 2020). Though perhaps extreme, it suggests that attention might be warranted for potentially offensive background audio as well. The expectations and standards in online learning are decidedly less rigid in higher education environments based on the assumption that these students are "adults." However, there were still instances of student disruption in lectures and online events where virtual audiences were subjected to deprecating outbursts and racial tirades (Brockington, 2022; Parkey, 2022).

Cellphone recordings of incidents at middle and high schools have been reported for years; however, few people expect or prepare for what they do in the face-to-face classroom or the online classroom to be brought to the forum of public opinion. While there are no surefire ways to eliminate all risk in a climate fraught with widespread feelings of offense, curating easily identifiable information about self and immediate family can help manage targetable points of impact for situations that suddenly trigger the spotlight of public scrutiny and opinion.

Communication Challenges in the Digital Environment

The pandemic forced most daily communication into an exclusively digital mode, and that social routine seems slow to revert. Conversations that were once in-person continue to take place in a digital realm out of convenience, but also where they can be stored and recalled ad nauseum. Once viewed by a select but relevant few, instructions, written examples, and educator syllabi are stored and shared indiscriminately across the web. Now that digitized documents are the norm rather than the exception, they are more susceptible to mass distribution. Despite differing opinions, personal growth and the evolution of views on political and social issues through conversation became stifled as digital communities formed echo chambers and introduced the fear of online harassment. The ease of taking screenshots and forwarding them only served to exacerbate miscommunication and strife. According to De Zwart et al. (2010), "(t)he very purpose of social networking sites, which is to lower the barriers to social communications, creates risks associated with uninhibited communications." Educators learned to operate on the assumption that their lectures and conversations were being recorded and possibly shared on social channels, which led to guarded speech and suppressed opinions (Redden, 2021). Teachers were also unable to make quick visual assessments of student understanding because either student cameras were off, or students were distracted by background activity.

Another challenge of online learning was that decisions regarding technology platforms and access were made at the administrative level, bypassing the input of the teachers who would be using them. Administrators often selected platforms based on user-friendliness and affordability rather than factoring in ease of use for room management or digital protection for students learning. Because online heckling was a completely new phenomenon for educators, administrators also failed to consider instructors teaching with an unintended audience. In addition to learning to use each unfamiliar platform, educators were simultaneously troubleshooting connectivity issues with students, trying to teach the curriculum in the least disruptive manner, and combating online aggression, leading teachers to unexpectedly lose classroom privacy and autonomy.

Community Engagement Expectations

Educators face unique complications when trying to minimize their level of personal exposure given that their job roles often include an expectation of community involvement. These mandatory but secondary responsibilities vary depending on the institution but can include coaching, club advising, and representing the school at various events. This exposure broadens at the college level with the additional expectations of grant-writing to fund research, scholarly publication, and presentation of the resulting research. In many cases, this is not optional because it is an evaluation requirement. This magnifies exposure beyond classroom walls and translates into broader recognition among those not directly tied to the instructor's primary responsibilities, which can include parents, students, boosters, alumni, and general community members at large. Many US schools have one or more social media accounts used for marketing and promotion of the organization. Administrators often use social media posts to share events, awards, and school news with the local community. To protect the identities of minors in those posts, school employees are often pictured and fully identified while leaving the students pictured unnamed. Teachers and staff receive neither compensation nor the same privacy protection as students when their images and reputations are used for marketing to increase school attendance and funding. These social media posts, however, expose identifying information about educators and increase their levels of visibility to the public.

Areas of Vulnerability

Understanding specific areas of vulnerability is the first step to regaining control of an educator's personal and professional identity. No two people will be juggling the same issues and scrutiny; therefore, carefully assessing their personal threat level is essential. Because outside influences inform some of these susceptibilities, they need to be revisited regularly and reflected on based on the current situation and the expectations of the educator's community. Engaging in thoughtful conversations with trusted colleagues can also be helpful. These conversations will allow opportunities to provide feedback on potential risks and to rethink threats to personal online identities. The pendulum of concern and scrutiny is ever-changing. Making informed and thoughtful decisions about how to actively engage in both personal and professional life provides the best protection to educators at any level or in any teaching environment.

Personal Identifiable Information

The authors define vulnerabilities as areas that leak Personal Identifiable Information (PII) about the educator. This information provides easy access for those looking to target or bully someone when there is disagreement or a difference of opinion. According to the National Institute of Standards and Technology (2022), PII is "information which can be used to distinguish or trace the identity of an individual alone or when combined with other personal or identifying information which is linked or linkable to a specific individual." Single bits of information can be aggregated to create a complex digital biography of an individual. As Solove (2004) explains "digital biographies greatly increase our vulnerability to a variety of dangers" (p. 146). PII includes name, address, date and/or place of birth, email address, telephone number, driver's license number, social security number, banking information, place of employment, names of family members, and any other information that can be used to identify and trace back to a specific educator.

Primary vulnerability is information that one unknowingly hemorrhages in various spheres of life, particularly where there is overlap. Educators should consider what goes online about their personal interests/passions and professional work/identity. It is easier than ever to connect those seemingly unconnected dots and merge those worlds without consent – often by people unknown to the target. With over 500 unique data brokers selling public record information online to anyone who pays, millions of records can now be searched with the click of a mouse button (Solove, 2020). For example, simple information such as name, birth date, and county of residence can be used to easily gain access to a person's home address on the voter registration website of most states. Unfortunately, this information is quite often disclosed voluntarily within social media posts and responses. Even if educators intentionally avoid disclosing personal data, friends and family may also inadvertently add to PII leakage by tagging and posting on the educator's social media accounts. The vulnerability intensifies when personal connections blend with friendly professional connections, so it is important that one reflects on how much risk they are willing to accept.

Information Institutions Must Share When Requested

Educators in public institutions are subject to both the federal Freedom of Information Act (FOIA) as well as the Open Records Act of the state in which they reside. These acts grant members of the public access to government documents upon request with a few well-defined restrictions, such as materials that involve law enforcement or national security. This includes requests for documentation or records involved in the daily work of state and federal employees. Most, if not all, of what an educator produces or uses at work is subject to disclosure to any member of the public who inquires. This includes emails, saved computer files, all hiring documents, employment evaluations, and more. Educators would be wise to take some time to locate and familiarize themselves with FOIA regulations found in their employee and institutional handbooks. Particular attention should be paid to social media use and electronic communication guidelines. Also, First Amendment law should be reviewed as it pertains to individual educators. These laws differ by state, district, and employer, so instructors should educate themselves within their own jurisdictions when outlining a privacy plan.

Information Voluntarily Disclosed

Educators should make time to identify how the deliverables in their work-life lead directly back to personal identifiable information about them or their families. Where can educators make autonomous decisions about their digital work identity and work product? They should make an appointment with someone from their human resources (HR) and/or information technology (IT) departments to discuss privacy options. Depending on the situation, it may be helpful if they can select the name used to identify them in work settings such as email, department directories, and so on. Also, educators should ask if it is possible to opt in or out of specific communication channels (e.g., Slack, LinkedIn Edu). Thought should be given to what information is shared about the educator in professional spheres, including conference proceedings, membership directories, committee minutes, and recordings. Items like these are often stored "in the cloud" and accessible to anyone motivated enough to search. Consider removing extraneous personal information from the professional biographies submitted for conference introductions and published works. Personal information such as names of immediate family, city of residence, or affiliations with specific community groups or organizations could be used to build a public attack or target the educator.

The information that is shared within an educator's personal activity groups and social media should be reviewed. Many personal and social connections triangulate with social media identities to create a treasure trove of critical personal identifiable information. This accidental hemorrhage of contact information, close acquaintances, and organizational affiliations can spill into one's professional life. When evaluating susceptibility between personal and professional personas, consider potential areas of unforeseen impact, such as membership directories for places of worship, homeowner associations, community clubs, groups, political organizations, and classroom directories. Think about what information is necessary versus information that may be considered oversharing. Use work emails and contact information exclusively for professional communication and rely on a separate personal email for family and non-work communication. Blending the two for convenience could lead to a massive headache later if the educator finds themself too reachable or needs to control damage and access to their online communication channels.

Ounce of Prevention

Given the already hectic schedule of the educator, taking steps to eliminate privacy vulnerabilities can be easily pushed to the back burner. Some educators may think, "there's no rush because it's unlikely to happen to me; my classes don't contain objectionable or controversial content," or "the time and effort to do all that seems exhausting." Privacy strategies and tools take some time but less time than it takes to deal with ransomware, identity theft, or doxing (A. Macrina, Library Freedom Project, personal communication, April 5, 2022). Being proactive will limit damage and prevent the need to scramble for new credentials and contact pathways.

There is no perfect plan to guarantee educator privacy and safety. There are, however, steps one can take to remove or mask personal identifiable information that is exposed and accessible to the public. As educators, the opportunities to accidentally hemorrhage basic personal information are numerous. Protecting family and professional identity requires a tremendous amount of diligence. To mitigate the damage caused by a doxing attempt, educators should consider the following:

- **masked phone numbers:** Use services that allow existing phone numbers to appear as masked numbers to non-family members. The masked number rings directly to the established number while avoiding potential abuse because masked numbers can be changed year over year or as needed with no impact on the established number.
- **email addresses:** When communicating with students or parents outside of a work email address, consider creating a separate email and using email forwarding services and/or secure email services rather than a personal email account. This gives more control over what is received and reduces the chances the personal address will need to be closed out because of abuse or hacking.
- **commercial mail receiving agencies (CRMA):** Use a CMRA (such as a UPS store) to receive mail off-site rather than a home address. This ensures that the address, displayed as a street address instead of a post office box number, will now be listed with data brokers rather than the geographical address of a physical home.
- **social media:** Create separate social media accounts for professional use. Review all personal social media accounts, remove unnecessary PII and lock down security and privacy settings.
- **information removal services:** Expedite the removal of personal information online by using an information removal service like DeleteMe to clean PII from social media accounts and/or data brokers.

Worst Case Scenarios

The authors would be remiss in failing to address the elevated dangers that exist in academia. Though these cases are less common than the bullying reported at the K-12 level, the professional impact can be devastating. With political polarization surging, there has been a sudden push to target college educators, their research, and course content. These attacks can get so personal that even physical appearance and racial slurs are fair game (Ferber, 2018). Unfortunately, the result is often the utter destruction of the accused's employment possibilities, academic credibility, and perceived moral character. Instructors are witnessing damage to reputations that they have built through the years "class by class and publication by publication" (Professor Washington-Hicks, personal communication, April 15, 2022).

Most of these harassment cases are instigated by groups that proactively search for targets connected to issues that conflict with their own worldviews, such as Critical Race Theory or Trans rights. A common misconception is that targeted harassment can be attributed entirely to overly zealous conservative groups. Political affiliation does not determine individual affect towards targeted harassment. Searches for victims often manifest in the formation of organizations devoted to identifying and targeting instructors whose teaching philosophies do not align with the groups' socio-political ideology. Three examples of such organizations are:

- **Puget Sound John Brown Gun Club:** In the text of their webpage, this left-wing group describes itself as an "anti-fascist, anti-racist, pro-worker community defense organization."
- **Professor Watchlist:** On their website, this organization boasts that their mission is "exposing and documenting college professors who discriminate against conservative students" by aggregating "instances of radical behavior among college professors."
- **Campus Reform:** The mission statement on this organization's website describes itself as "a conservative watchdog to the nation's higher education system."

When focused organizations target educators with opposing viewpoints, they trigger unnecessary fervor in individuals who can take retaliation to frightening extremes. This behavior can be found in both higher education and K-12 settings. Between 2016 and 2018, over 200 university professors were targeted as victims of online harassment based on their research topics, teaching, or things they posted on social media (Kamenetz, 2018). For instance, Professor Keeanga-Yamahtta Taylor of Princeton University faced retaliation after a news story profiled a commencement speech that she delivered at Hampshire College in 2017. Professor Taylor, who still has a profile on Professor Watchlist, was forced to cancel multiple public speaking appearances due to numerous death threats she received following an opinion piece broadcast by Fox News (Flaherty, 2017). After being harassed over a perceived link to Critical Race Theory, newly hired Cecelia Lewis declined an initial job offer for a DEI administrative role in favor of a position with another nearby county. The same group of rural North Georgia community members tracked her to the new position and continued their intimidation which led to a second resignation and an out-of-state relocation (Carr, 2022). Similarly, but on the other end of the political spectrum, fourth-grade teacher Kristine Hostetter was suspended for over a year after a video was posted of her in attendance at the January 6th Capitol protests (Rosenberg, 2021). Although the common consensus was that she was an excellent teacher, her neighbors and students tracked her family's social media accounts and signed multiple petitions to have her removed from her teaching position because they disagreed with her political views.

Facing a New Reality

Harassment of educators has become disturbingly common since the start of the pandemic. According to the American Psychological Association (2022), 50% of the teachers responding to a survey on educator harassment expressed a plan or desire to quit the profession due to violence and threats received during the pandemic. Regardless of where they fall on the political spectrum, teachers endure most of the blame for an education system they did not create and never controlled. The teaching profession is a historically low-paying position of service held to both internal (institutional) account and external (the public) account. Trying to balance and appease these two groups while simultaneously adhering to the ideals of the profession can be demanding even in the best of environments. Accepting the risks required of and the vulnerabilities inflicted upon educators who choose to remain in the field to support students is rewarded with threats, bullying, and reputational damage. Because the wave of public opinion is quick to change course, educators can no longer predict what will attract unwanted and unwarranted scrutiny. Educators at all levels should complete a diligent review of their existing PII and the overlap between their personal and professional identities and follow up with regular re-checks to protect themselves from the worst-case scenarios of being targeted online.

Recommended Reading

Unfortunately, there is no single, definitive privacy source that fits the needs of everyone. The authors offer a non-exhaustive recommended reading list as sources to begin learning more about digital privacy. We further suggest that the reader follow up with their own research and determine the privacy model that fits their individual needs.

Freedom of Information Act (FOIA)
The United States Department of Justice publishes this website about the FOIA. It walks the user through the process of filing a FOIA request and has an extensive frequently asked questions section. https://www.foia.gov/

Library Freedom Project
This organization teaches librarians about surveillance threats, privacy rights, and digital tools to thwart surveillance. The resources section provides posters, bookmarks, presentations, and more to help anyone interested in teaching and promoting privacy. https://libraryfreedom.org/

Firewalls Don't Stop Dragons (Corey Parker)
The latest (2020) version of this book is an excellent reference for anyone who needs to protect their digital identity. The author covers everything from passwords to mobile phone security in everyday language. The author also has a great blog by the same name that he keeps current with weekly posts. Book ISBN: 978-1484261880 BLOG: https://firewallsdontstopdragons.com/

Electronic Frontier Foundation (EFF)
This non-profit organization defends civil liberties in the digital world by championing user privacy, free expression, and innovation. https://www.eff.org/

Corporate Surveillance in Everyday Life
This site is published by Cracked Labs, a non-profit organization that investigates the socio-cultural impacts of information technology. The site does a thorough job of defining data brokers and explaining the impact of digital surveillance on the average consumer. The infographics alone are worth the effort of viewing the site. https://crackedlabs.org/en/corporate-surveillance/#4

Surveillance Self-Defense (EFF)
An expert guide, published by EFF, detailing how to protect family and friends from online spying. https://ssd.eff.org/

References

Agrawal, N. (2020, August 28). T-shirt ignites threats against teacher. *Los Angeles Times*.

American Psychological Association. (2022, March 17). *Teachers, other school personnel, experience violence, threats, harassment during pandemic* [Press release]. https://www.apa.org/news/press/releases/2022/03/school-staff-violence-pandemic

Armental, M. (2020, October 12). PC demand during pandemic fuels strongest U.S. market growth in a decade; Chromebook shipments surged about 90% in the third quarter because of distance learning. *Wall Street Journal (Online)*. https://www.wsj.com/articles/pc-demand-during-pandemic-drives-strongest-u-s-market-growth-in-a-decade-11602537956

Associated Press. (2020, September 26). Social media takes COVID-19 shaming to new levels. *Los Angeles Times*. https://www.latimes.com/world-nation/story/2020-09-27/social-media-and-covid-shaming-fighting-a-toxic-combination

Brockington, A. (2022, February 3). *Olivia Munn opens up about racist Zoombombing during AAPI meeting: 'It was jarring.'* NBC News.

https://www.nbcnews.com/news/asian-america/olivia-munn-opens-racist-zoombombing-aapi-meeting-was-jarring-rcna14741

Campus Reform. (2022). *Mission.* https://www.campusreform.org/about

Carr, Nicole. (2022, June 16). *White parents rallied to chase a Black educator out of town: Then they followed her to the next one.* Frontline–PBS. https://www.pbs.org/wgbh/frontline/article/crt-georgia-schools

Coghill, A. (2022, February 4). *Glenn Youngkin set up a tip line to snitch on teachers: It's only gotten weirder since.* Mother Jones. https://www.motherjones.com/politics/2022/02/glenn-ngkin-virginia-tip-line-crt-teachers/

De Zwart, M., Henderson, M., Phillips, M., & Lindsay, D. (2010). 'I like, stalk them on Facebook': Teachers' 'privacy' and the risks of social networking sites. *2010 IEEE International Symposium on Technology and Society, 2010*, 319–326. https://doi.org/10.1109/ISTAS.2010.5514624

Faverio, M. (2022, January 13). *Share of those 65 and older who are tech users has grown in the past decade.* Pew Research Center. https://www.pewresearch.org/fact-tank/2022/01/13/share-of-those-65-and-older-who-are-tech-users-has-grown-in-the-past-decade/

Ferber, A. (2018). 'Are you willing to die for this work?': Public targeted online harassment in higher education. *Gender & Society, 32*(3), 301–320. https://doi.org/10.1177/0891243218766831

Flaherty, C. (2017, June 1). *Concession to violent intimidation.* Inside Higher Ed. https://www.insidehighered.com/news/2017/06/01/princeton-professor-who-criticized-trump-cancels-events-saying-shes-received-death

Jargon, J. (2020, October 27). Parents are the new remote-school zoom bombers; moms and dads fight the urge to overstep in virtual classrooms. To the chagrin of teachers, some can't help themselves. *Wall Street Journal (Online).* https://www.wsj.com/articles/parents-are-the-new-remote-school-zoom-bombers-11603800001

Kaffer, N. (2020, April 19). New norm becomes cold, quiet, less comforting: Gatherings of family and friends, once full of life, have succumbed to safety measures. *Detroit Free Press.*

Kamenetz, A. (2018, April 4). *Professors are targets in online culture wars: Some fight back.* NPR. https://www.npr.org/sections/ed/2018/04/04/590928008/professor-harassment

Major, C. H. (2015). *Teaching online: A guide to theory, research, and practice.* Johns Hopkins University Press. https://doi.org/10.1353/book.38784

McMahon, S. D., Anderman, E. M., Astor, R. A., Espelage, D. L., Martinez, A., Reddy, L. A., & Worrell, F. C. (2022). *Violence against educators and school personnel: Crisis during COVID*. Technical Report. American Psychological Association.

National Institute of Standards and Technology. (2022, March 10). *Glossary*. Computer Security Resource Center. https://csrc.nist.gov/glossary/term/PII

Parkey, A. (2002, February 1). Shift to virtual life gives rise to 'zoombombing' and racist outbursts. *Daily Iowan*. https://dailyiowan.com/2022/02/01/shift-to-virtual-life-gives-rise-to-zoombombing-and-racist-outbursts-online/

Pawlewicz, D. (2020). *Blaming teachers: Professionalization policies and the failure of reform in American history*. Rutgers University Press.

Pressman, A. (2021, January 11). PC sales have surged for at-home workers and learners during the pandemic. *Fortune*. https://fortune.com/2021/01/11/covid-computer-sales-lenovo-hp-dell-apple/

Professor Watchlist. (2022). *Exposing bias on campus*. https://professorwatchlist.org/

Redden, E. (2021, March 12). Georgetown professor fired for statements about Black students. *The Chronicle of Higher Education*. *https://www.insidehighered.com/news/2021/03/12/georgetown-terminates-law-professor-reprehensible-comments-about-black-students*

Rosenberg, M. (2021, April 10). A teacher marched to the capital, when she got home, the fight began. *New York Times*. https://www.nytimes.com/2021/04/10/us/politics/kristine-hostetter-capitol.html

Solove, D. (2004). *The digital person: Technology and privacy in the information age*. New York University Press.

Weale, S. (2022, April 17). Parents targeting teachers with 'aggressive' emails since Covid outbreak. *The Guardian*. https://www.theguardian.com/education/2022/apr/17/parents-targeting-teachers-with-aggressive-emails-since-covid-outbreak

Online Harassment in Elementary Schools

Rebecca Taylor

Technology has long been integrated into the educational world, but it has become a necessity since the beginning of the COVID-19 pandemic. Educators have devoted a lot of time to teaching students how to use educational technology for their schoolwork, but did they properly prepare students for how to socialize through technology? This question is emphasized when considering cyberbullying. Cyberbullying incidents have increased as the use of technology is increasing among younger and younger children. It impacts all students involved and can continue to impact these students after their school career ends; therefore, it is important for schools and parents to work together to provide a united front against cyberbullying. Teachers may ask to what extent our responsibility goes to raise awareness of and prevent cyberbullying. Answering this question begins by studying existing research and prevention strategies.

Research summarized by the PACER Center (2020) indicates that online harassment has become increasingly more common in elementary schools, yet for the most part, students have not received more information about it. While using educational technology, students have been caught sending hateful messages in chat boxes, hacking other students' accounts to get them in trouble, and impersonating adults online. Students can create false profiles and hide their identities, enabling them to say anything to others without facing a consequence (Donegan, 2012). If they are using social media, it can be extremely difficult for accounts to be verified, as this only works for well-known celebrities (Karmaker & Das, 2020). It is important for educators and communities to ask themselves, how do we fix this behavior or prevent it from happening altogether?

This chapter will explore various bullying intervention and prevention strategies, with a particular emphasis on cyberbullying and challenges related to the increase in online learning during the COVID-19 pandemic.

Elementary School Bullying

Through guidance lessons, elementary school students are frequently taught the importance of being kind to others and the hurtful damage caused by bullying. These lessons are important because elementary students often do not understand the emotional effects of their words. While these lessons have been geared towards in-person situations and learning, the shift toward remote learning during the COVID-19 pandemic demands the inclusion of these lessons as they apply to online learning. Throughout the pandemic, students have been moving in and out of quarantine and have been expected to interact with teachers and peers through online platforms. In elementary school, many students do not understand the effects their harsh words can have during face-to-face interactions, so it is even harder for them to identify their own behavior as bullying when they are able to hide behind a computer screen.

For adults, identifying bullying before it becomes a problem is difficult. According to Fienberg and Robey (2009), students are often reluctant to share bullying issues with teachers or parents "because they are emotionally traumatized, think it is their fault, fear retribution, or worry that their online activities or cell phone use will be restricted" (p. 2). While there are technological aids to identify and stop bullying, they only really work for school-issued equipment. For example, there are platforms for middle and high schools–such as Veyon, Kickidler, Classroom Spy, and NetSupport–that allow educators to see the computer screens of their students, and elementary schools tend to block certain websites to prevent access to applications like YouTube (Lynch, 2018). However, these measures do not fully prevent students from bullying, accessing inappropriate websites, and completing unrelated school activities, especially since schools have no control over students' personal devices at home.

One difficult issue is that the school environment is fluid: students move through different classrooms each day, different grade levels with different teachers each year, and eventually to different schools. Each educator or school only has a limited time to stop or slow the bullying that is occurring. In addition, students are at home in different types of life situations, which can influence how they behave in all other environments. Students come from a range of family types, including traditional, single parent, blended, or grandparent families. Despite all the different variables involved, research does show that talking about bullying and how it is not allowed or tolerated can prevent students from partaking in bullying behaviors (Donegan, 2012).

Differences in Cyberbullying

Bullying is a common issue in many schools across the world and can take various forms including verbal threats, physical assaults, and online insults

(Storey & Slaby, 2013). Cyberbullying is defined as the transmission of harmful or cruel text or images online using internet platforms, such as social media or online forums, or digital devices using direct messages. It can be presented as flaming, harassment, stalking, impersonation, gossip, outing, or exclusion (Feinberg & Robey, 2009). Cyberbullying can often cause more damage than traditional bullying due to the unlimited access to people online and a lack of adult supervision. Paek et al. (2022) discuss how a lack of parental supervision is a noticeable predictor for online victimization. This type of bullying does not depend on environmental influence or motivation as the perpetrator can be anyone, even someone the student does not know. In many cases, there is no specific reason behind cyberbullying other than opportunity (Notar et al., 2013). Cyberbullying can render specific negative impacts on students due to the added vulnerability of the online environment. Donegan (2012) explains that "online publication of personal information is dangerous because it allows many people to see a side of a person more often kept private in a face-to-face interaction. This vulnerability puts many teens in a position, as either the victim or active offender, to partake in cyberbullying actions" (p. 35). Elementary-age children are particularly unaware of their online vulnerability and how much more of themselves may be exposed online than through in-person interactions (Donegan, 2012). These younger students may also be unaware of how many people have access to their online content.

Incidents of cyberbullying among students have been increasing due to the COVID-19 pandemic (Karmaker & Das, 2020). During this time, schools went completely online for virtual learning and younger students, many for the first time, were taking technology home to use. Students now had access to communication with other students, and several students have used the technology to bully others. For instance, Karmakar and Das analyzed public tweets from January to July 2020 and found "a clear telling effect of COVID-19 on worsening cyberbullying incidents as reported and discussed through tweets" (2020, p. 2). Kee et al. (2022) also reported an increased use of social media during COVID-19 leading to an increased risk of cyberbullying. This study found that the COVID-19 pandemic caused a distinct increase in cyberbullying among the youth surveyed.

Conversely, Mkhize and Gopal (2021) found evidence of more children and youth becoming involved in social media during this time but not concrete evidence of an increase in cyberbullying. However, more exposure to online communication means more risk of cyberbullying. Patchin (2021) found that face-to-face bullying significantly dropped while cyberbullying remained consistent. Despite the different conclusions as to whether cyberbullying remained stable or increased, there is agreement among these studies that more children are being exposed to online experiences, and children need to learn about appropriate online communication and how to handle cyberbullying should it occur.

Social Media's Effect on Bullying

Social media use continues to grow, and many elementary-age children are using it to engage in communication. Several young students are on Facebook, Twitter, Instagram, Snapchat, or TikTok. While using these accounts, students witness adult language and content because most of the online platforms do not have a way to select age-appropriate content. Although unrestricted social media sites typically have minimum age requirements, it is easy for younger users to manipulate this by claiming to be an older age (Pasquale et al., 2020).

These students are subsequently exposed to cyberbullying and inappropriate content through these platforms. In fact, it could be the presence of these younger children that contributes to a more hostile online environment, as several researchers have noted that elementary students display cyberbullying behaviors more often than older students (Biggs et al., 2010).

While the age limit on some of the platforms is not stopping all cyberbullying, it can be an extra defense to slow it down. Oftentimes, students are not being supervised while they are on these platforms, and parents are unaware of how exposed their children are online. These online platforms can make bullying easier and affect school behavior if they are not properly monitored. For example, TikTok challenges that have been promoted in the 2021-2022 school year have involved vandalizing schools, slapping teachers, and bringing weapons (Walie, 2021). How do schools and educators teach the negative effects and repercussions on the topic of online harassment when this type of behavior is encouraged on the social media students are accessing at home? During the COVID-19 pandemic, students were accessing more social media while parents were still busy working, either in person or online, and not able to constantly supervise their children (Agostinelli et al., 2022).

Social media makes it easier for bullying to happen "because airing one's opinion, sending hate mails, recording videos and uploading photos are easier, [so] ridiculing someone is also easier, especially [on] social media" (Santiago, 2015, p. 96). It is also important for students to know that what is written on the internet can rarely be erased, and it can be shared and accessed by anyone. This relationship between social media and cyberbullying is only getting worse, as social media is more frequently used for negative influence.

Lasting Effects of Bullying

Cyberbullying has several negative effects on students of any age, and they can carry these effects with them into adulthood. The first negative effect is more violence and more bullying. Children who are the victims of bullying often bully other students as a result (Rigby & Slee, 1999). Studies have also shown that a large percentage of bullying victims feel vengeful afterward (Donegan, 2012).

In addition to unleashing violent anger on others, many children have committed suicide after being harassed online (Santiago, 2015). Lack of adult supervision and easy access to social media make it easy for students to go from victim to bully, with severe results. Paek et al. (2022) find that parental supervision can help reduce cyberbullying victimization and aftereffects, such as mental health consequences.

When thinking of the effects of cyberbullying on schools today, people often think of middle school and high school students who have more access to technology. According to the National Center for Education Statistics (2022), approximately 16% of 9th to 12th graders had experienced cyberbullying in the recent school year. Adding in middle schoolers, the Pew Research Center found that more than half of 13–17-year-olds (59%) had been the target of some form of cyberbullying, with offensive name-calling and rumor-spreading topping the list of most common incidents (Anderson, 2018). And on the youngest end of the scale, one in five tweens (20.9%), defined as ages 9-12, has been a victim, perpetrator, or witness to cyberbullying (Patchin & Hinduja, 2020).

While the emotional effects of cyberbullying vary by individual, they also tend to vary by age group. While some older students may respond with frustration and have a drive to prove themselves to competitors, elementary students more often become sad, occasionally leading them into depression or anxiety (Donegan, 2012; Doumas & Midgett, 2020). Students who are victims of bullying can begin to feel withdrawn from school. Their self-esteem drops the more they are bullied, and they can feel like they no longer belong at school. Students can become dejected while at school and this can affect the friendships victims have with others (Holder & Coleman, 2008; Torres et al., 2019). They can also experience mental health concerns, including depression and anxiety (Doumas & Midgett, 2020). Depression and anxiety can impact these students and their relationships with others long after they are out of school and can manifest in many adult traits such as shyness, low self-esteem, and withdrawal (DePaolis & Williford, 2015).

The effects of cyberbullying can also lead to academic concerns for victims throughout their school careers. This is partly because students who are bullied often try to miss school to avoid being a victim, which naturally causes a decline in academic performance (Rivers, 2000). These repeated absences can affect student motivation and commitment to learning, despite any natural academic abilities and previous achievements. Peled (2019) recognizes the importance of motivation and commitment by saying, "motivation to learn, taking actions to meet academic demands, a clear sense of purpose, and a general satisfaction with the academic environment are also important components of the academic field" (p. 7).

As previously noted, bullying affects not only the victim but also the bully and bystanders. Rigby and Slee (1999) state that young people who bully as well as their victims are at an increased risk for suicidal thoughts, suicide attempts, and completed suicides. Bystanders of cyberbullying are also at a risk for mental health concerns, including "depressive symptoms and social anxiety over and above the effects of witnessing school bullying and bullying victimization" (Doumas & Midgett, 2021, p. 4) This could be due to students feeling more helpless when reporting cyberbullying because the perpetrator can be unknown, and the bullying could happen at any time (Feinberg & Robey, 2009). Therefore, the bully could be hiding their identity and posting harsh words towards the victim essentially 24 hours a day.

Intervention

Because of the pervasive use of technology both in and outside of school, it is nearly impossible to prevent cyberbullying from occurring altogether. Therefore, since so many students may one day experience cyberbullying, schools and parents need to be aware of intervention and treatment programs. The treatment process depends on the victim's mental health concerns, whether that is emotional, mental, or psychological. All victims could be treated for possible depression, anxiety, and self-esteem, as these are the more common outcomes (Doumas & Midgett, 2021).

One good intervention is teaching victims how to deal with cyberbullying should it occur in the future (Doumas & Midgett, 2021). For shy students who may become victims, they can be taught how to assertively use the word "no," while potential bystanders need to work on problem-solving skills to stand up for the victim and help stop the incident without being aggressive (Storey & Slaby, 2013; Thornberg et al., 2012). Those who display bullying behaviors need practice with social skills, such as empathy. There are several skills to help students act in appropriate ways as victims or bystanders that can decrease depression and anxiety after cyberbullying. Students need to have options on how to respond to bullying incidents in the future. The number of students experiencing or witnessing cyberbullying is exceptionally high, so it would be a disservice to students to not prepare them.

School guidance counselors are a natural resource for teaching social skills and implementing anti-bullying programs. One way that counselors can help students is through what might be called a "lunch bunch" (Woolf, 2022). The purpose is to have lunch with a small group of students and help them grow a friendship. This time can be used for small group intervention skills, which can be helpful for bullies, victims, and bystanders. These learned skills are also helpful for victims with social anxiety and students who have started to depersonalize from school, as interactions with smaller groups can help these students build confidence. This may help students break away from the vicious exhausting depression cycle or provide students with a group to share and create friendships.

Guidance counselors can use their "lunch bunch" groups to deploy two main strategies to help intervene with cyberbullying. The first strategy is to implement a program that provides students with social skills and problem-solving behaviors. The skills are focused on improving poor social skills and developing interpersonal skills (Woolf, 2022). The second strategy is to provide skills for coping as a victim of bullying. These skills range from thinking positively to analyzing an issue. Both sets of skills are essential for all students because negative emotions can hinder academic achievement (Torres et al., 2019). Accordingly, parents should be informed of students who may have experienced cyberbullying and the effects it can have on their academic performance.

Extracurricular activities are another promising strategy to improve socialization among students. These activities can be highly encouraged for students, and in some schools, it may be mandatory for students to participate in at least one. The schools must be able to provide options and opportunities for these activities due to the socialization benefits. These activities can help victims of bullying feel more productive and more confident in their social abilities, and they can help the bullying perpetrators, too, by encouraging them to make new friends. As students are engaged in more constructive activities, their negative online encounters may decrease (Santiago, 2015). These beneficial activities can be extracurriculars taking place both in and outside of schools.

Last, one final intervention is simply talking about and making students aware of cyberbullying. Students need relief from the emotional impacts of this issue; however, since they are unlikely to find relief from mentors, they are often left feeling helpless (Donegan, 2012; Storey & Slaby, 2013). The more schools, parents, and communities discuss the harm caused by bullying, the more comfortable students may feel coming forward. Students need a safe space, free of threats, where they feel able to seek help from all adults. This can come from a consistent intervention program, in which the students receive time to talk to an adult, work through the incident, and discuss how to fix it in a safe environment (Caines, 2021). As the school environment becomes more caring and safer, educators, parents, and students can work together to increase student success (Coloroso, 2016).

Prevention

Donegan (2012) notes that the bullying and cyberbullying problem is so scary because it can never fully be stopped, in part "due to how deep seeded [the problems] have become in our competitive society" (2012, p. 39). Such competition is a part of all stages of American life, from college applications to the corporate world. This is the harsh reality, and the use of social media adds to this type of pressure, making bullying possible worldwide, day or night (Doumas & Midgett, 2021). Pre-teens and teenagers are especially susceptible to this because of the vulnerability they show online.

Schools need to work harder to end cyberbullying, especially during times of greater online activity, such as the period since COVID-19 began. So how is this possible? Providing intervention and prevention policies can be a step towards making the idea of a threat-free school a reality. Donegan (2012) gives educators hope by discussing how prevention programs are becoming more effective as educators are learning specific bullying tactics and the reasoning behind the bullying.

One way to promote prevention is by utilizing the guidance counselor. These counselors are an asset due to their experience with providing social skills, building relationships, helping students feel safe, and their knowledge of bullying. These experts play a major role in the prevention, intervention, and after-effects of bullying. Doumas and Midgett (2021) point out that school personnel can be essential for providing information on the importance of education for witnessing cyberbullying, bystander behavior, and how to report this type of bullying. It is equally important for these school personnel to inform parents about bullying for victims and witnesses (p. 632). This stresses the importance of using counselors and letting them work with students for the prevention of cyberbullying issues. These findings also show schools and parents how important it is to talk about cyberbullying and its negative impact on all parties involved (Health Resources & Service Administration, 2022).

Guidance counselors can provide information on the effects of academic performance and mental health. They can also share the school prevention plan and state bullying/harassment laws, which provide procedures and measures for a school district to address bullying incidents and how to report them (Assistant Secretary for Public Affairs, 2021). Donegan (2012) shows educators how helpful this is by saying, "if American communities and schools address the issue with a clear preventative program that keeps each level of prohibition on the same page, children will in turn receive a consistent message from a young age, which will presumably resonate effectively" (p. 39). This echoes the importance of sharing information and providing consistency on cyberbullying so that students can begin to understand consequences, the impact bullying has on other children, and that the school is on a team to help students feel safe.

The last way to utilize guidance counselors is by letting them implement programs for preventative measures and teaching lessons on this topic. Table 1 shows an example of guidance lessons taught for the entire school year in the author's elementary school in Tennessee.

Table 1
Elementary Guidance Lessons in a Tennessee School

Kindergarten and 1st	2nd and 3rd	4th and 5th
• Bullying • Tattling • Accepting Others • Trustworthiness • Good touch/bad touch	• Bullying • Citizenship • Coping Skills • Prioritizing Work over Play • Study Skills	• Bullying • Problem Solving • Kindness Matters • SMART Goals • Study Skills • How to do Homework • How actions / choices lead to consequences

Notice that cyberbullying is not explicitly addressed, possibly due to the assumption that students in middle and high school are facing more cyberbullying than younger students. However, as mentioned earlier, the number of students online in elementary school is growing, and they may even engage in more cyberbullying than older students with a higher impact (Biggs et al., 2010; Chen & Cheng, 2017). Therefore, lessons about cyberbullying should begin sooner. The first lesson on bullying is strictly focused on in-person, face-to-face instances. Cyberbullying, along with traditional bullying, needs an independent lesson at the beginning of the year in grades third through fifth, due to cyberbullying being more prevalent in these grades (Wilkey Oh, 2019). Coping skills, problem-solving, and how actions lead to consequences would integrate into these lessons.

The final prevention tactic is from Karmakar and Das (2020), who discuss possible defenses against cyberbullying as technical mitigation techniques, organizational policies, and user perspective. Technical mitigation techniques seek to detect cyberbullying and intervene before it goes too far. One possible solution is a dashboard teachers can use to monitor interactions on school devices using natural language processing. Keywords and phrases would be flagged as potential bullying incidents. Organizational policies focus on social media and how to protect the victims of cyberbullying on these sites. While it is tricky to police behavior that happens off school grounds and platforms, clear school policies can make it possible to provide consequences, and even bring in law enforcement backup when necessary. User perspective involves research on the perpetrators of cyberbullying. This emphasizes finding the motivational reason behind the bullying to find prevention strategies.

Conclusion

In closing, it appears that the rise in online schooling caused by COVID-19 correlates to an increase in cyberbullying. Cyberbullying is known to have negative emotional and social impacts on countless children, and those effects can persist long-term. This problem is growing as social media use increases among younger students. The use of technology in and out of the school environment is constantly growing as well, thus leading to more opportunities for students to become involved in cyberbullying as bullies, victims, and bystanders. Cyberbullying prevention must be a priority due to the possibility of students having mental health concerns, including depression and suicidal thoughts. Santiago (2015) reaffirms this point saying, "children might take this behavior with them even after they leave schools, so teachers should apply policies that will improve the safety and happiness of the students, and to show bullies that any of these acts are unacceptable in schools" (p. 98). If parents and the community present a united front, students will have support once they leave the school atmosphere. Ultimately, schools should be a safe space for students to grow in knowledge and maturity, and this is achievable through utilizing parent help, guidance counselors, and all school personnel in the fight to keep schools free of bullying.

References

Agostinelli, F., Doepke, M., Sorrenti, G., & Zilibotti, F. (2022). When the great equalizer shuts down: Schools, peers, and parents in pandemic times. *Journal of Public Economics, 206*. https://doi.org/10.1016/j.jpubeco.2021.104574

Aizenkot, D. (2020). Social networking and online self-disclosure as predictors of cyberbullying victimization among children and youth. *Children and Youth Services Review, 119*. https://doi.org/10.1016/j.childyouth.2020.105695

Anderson, M. (2018). *A majority of teens have experienced some form of cyberbullying*. Pew Research Center. https://www.pewresearch.org/internet/2018/09/27/a-majority-of-teens-have-experienced-some-form-of-cyberbullying/

Assistant Secretary for Public Affairs (ASPA). (2021, November 2). *Tennessee anti-bullying laws & policies*. StopBullying.gov. Retrieved September 4, 2022, from https://www.stopbullying.gov/resources/laws/tennessee

Biggs, B. K., Vernberg, E., Little, T. D., Dill, E. J., Fonagy, P., & Twemlow, S. W. (2010). Peer victimization trajectories and their association with children's affect in late elementary school. *International Journal of Behavioral Development, 34*(2), 136–146. https://doi.org/10.1177/0165025409348560

Caines, A. (2021). Keeping school learning environments safe from bullying. *BU Journal of Graduate Studies in Education, 13*(3), 26–30.

Chen, L., & Cheng, Y. (2017). Perceived severity of cyberbullying behaviour: Differences between genders, grades and participant roles. *Educational

Psychology, 37(5), 599–610. https://doi.org/10.1080/01443410.2016.1202898

Coloroso, B. (2016). *The bully, the bullied, and the not-so-innocent bystander: From preschool to high school and beyond: Breaking the cycle of violence and creating more deeply caring communities* (6th ed.). William Morrow Paperbacks.

DePaolis, K., & Williford, A. (2015, June). The nature and prevalence of cyber victimization among elementary school children. *Child & Youth Care Forum, 44,* 377-393.

Donegan, R. (2012). Bullying and cyberbullying: History, statistics, law, prevention and analysis. *The Elon Journal of Undergraduate Research in Communications, 3*(1), 33-42.

Doumas, D. M., & Midgett, A. (2021). The association between witnessing cyberbullying and depressive symptoms and social anxiety among elementary school students. *Psychology in the Schools, 58*(3), 622–637. https://doi.org/10.1002/pits.22467

Feinberg, T., & Robey, N. (2009). Cyberbullying: intervention and prevention strategies. *National Association of School Psychologists, 38,* 1–4.

Health Resources and Services Administration (HRSA). (2022, March 29). *Stop bullying home page.* StopBullying.gov. Retrieved July 13, 2022, from https://www.stopbullying.gov/

Holder, M. D., & Coleman, B. (2008). The contribution of temperament, popularity, and physical appearance to children's happiness. *Journal of Happiness Studies, 9*(2), 279–302. https://psycnet.apa.org/doi/10.1007/s10902-007-9052-7

Karmakar, S., & Das, S. (2020, November). Evaluating the impact of COVID-19 on cyberbullying through Bayesian trend analysis. *Proceedings of the European Interdisciplinary Cybersecurity Conference* (pp. 1-6). https://doi.org/10.1145/3424954.3424960

Kee, D. M. H., Al-Anesi, M. A. L., & Al-Anesi, S. A. . (2022). Cyberbullying on social media under the influence of COVID-19. *Global Business and Organizational Excellence, 41*(6), 11–22. https://doi.org/10.1002/joe.22175

Lynch, M. (2018, April 25). *How to use technology to prevent school bullying.* The Edvocate. Retrieved July 11, 2022, from https://www.theedadvocate.org/use-technology-prevent-school-bullying/

Miller, A., & Taylor, R. (2022). *Elementary guidance lessons in a Tennessee school* [Unpublished manuscript].

Mkhize, S., & Gopal, N. (2021, February). Cyberbullying perpetration: Children and youth at risk of victimization during COVID-19 Lockdown. *International Journal of Criminology and Sociology.*

National Center for Educational Statistics. (2022). Bullying at school and electronic bullying. *Condition of Education*. U.S. Department of Education, Institute of Education Sciences. Retrieved from https://nces.ed.gov/programs/coe/indicator/a10

Notar, C. E., Padgett, S., & Roden, J. (2013). Cyberbullying: A review of the literature. *Universal Journal of Educational Research, 1*(1), 1–9. http://doi.org/10.13189/ujer.2013.010101

PACER's National Bullying Prevention Center. (2020, November). *Bullying statistics*. Retrieved July 12, 2022, from https://www.pacer.org/bullying/info/stats.asp

Paek, S. Y., Lee, J., & Choi, Y. J. (2022). The impact of parental monitoring on cyberbullying victimization in the COVID-19 era. *Social Science Quarterly, 103*(2), 294–305. https://doi.org/10.1111/ssqu.13134

Pasquale, L., Zippo, P., Curley, C., O'Neill, B., & Mongiello, M. (2020). Digital age of consent and age verification: Can they protect children? *IEEE Software, 39*(3), 50–57. https://doi.org/10.1109/MS.2020.3044872

Patchin, J. W. (2021, September). *Bullying during the COVID-19 pandemic*. Cyberbullying Research Center. Retrieved September 4, 2022, from https://cyberbullying.org/bullying-during-the-covid-19-pandemic

Patchin, J. W., & Hinduja, S. (2020). *Tween cyberbullying in 2020*. Cyberbullying Research Center and Cartoon Network. Retrieved from: https://i.cartoonnetwork.com/stop-bullying/pdfs/CN_Stop_Bullying_Cyber_Bullying_Report_9.30.20.pdf

Peled, Y. (2019). Cyberbullying and its influence on academic, social, and emotional development of undergraduate students. *Heliyon, 5*(3). https://doi.org/10.1016/j.heliyon.2019.e01393

Rigby, K., & Slee, P. T. (1999). Suicidal ideation among adolescent school children, involvement in bully-victim problems, and perceived social support. *Suicide and Life-Threatening Behavior, 29*(2), 119–130.

Rivers, I. (2000). Social exclusion, absenteeism and sexual minority youth. *Support for Learning, 15*(1), 13–18. https://doi.org/10.1111/1467-9604.00136

Santiago, M. G. (2015). Anti-cyberbullying activities for elementary schools. In N. Naldoza (Ed.), *Contemporary issues in education: A compendium* (pp. 95-102). Polytechnic University of the Philippines.

Storey, K. & Slaby, R. (2013). *Eyes on bullying in early childhood*. Education Development Center.

Torres, C. E., D'Alessio, S. J., & Stolzenberg, L. (2020). The effect of social, verbal, physical, and cyberbullying victimization on academic performance. *Victims & Offenders*, *15*(2), 1–21. http://dx.doi.org/10.1080/15564886.2019.1681571

Walie, L. (2021, November 22). *'Devious licks' TikTok challenges sweep the nation, impacting schools everywhere*. The City Journals. Retrieved July 13, 2022, from https://www.mysugarhousejournal.com/2021/11/22/375777/-devious-licks-tiktok-challenges-sweep-the-nation-impacting-schools-everywhere

Wilkey Oh, E. (2019, March 15). *Teachers' essential guide to cyberbullying prevention*. Common Sense Education. Retrieved July 12, 2022, from https://www.commonsense.org/education/articles/teachers-essential-guide-to-cyberbullying-prevention

Woolf, N. (2022). *The lunch bunch: A small-group intervention for building social skills*. Panorama Education. Retrieved July 12, 2022, from https://www.panoramaed.com/blog/lunch-bunch-intervention

Ask What You Want; We Don't Know Who You Are: Live Chat, Library Anxiety, and Privacy in an Academic Library

Bridgette Sanders, Jon B. Moore, and Kimberly Looby

Virtual reference services have evolved over the last 20+ years and are now a significant part of many libraries. According to the Reference and User Services Association (RUSA), "Virtual reference is a service initiated electronically for which patrons employ technology to communicate with public services staff without being physically present. Communication channels used frequently in virtual reference include chat, videoconferencing, Voice-over-IP, co-browsing, e-mail, instant messaging, and text" (RUSA, 2017). Chat has been an effective means of contact by patrons to their library for many years. It allows patrons to ask questions quickly and efficiently without calling or speaking with someone. This may reduce patron uneasiness and library anxiety, which is the "uncomfortable feeling or emotional disposition experienced in a library setting that has cognitive, affective, physiological, and behavioral ramifications" (Jiao & Onwuegbuzie, 1997, p. 372).

In the literature of library science, one recognizes the interplay between embarrassment and asking for help as a well-known feature of library anxiety. Librarians often consider live chat as one way of alleviating library anxiety because of its remote access and enhanced anonymity (Brown, 2011; Fagan & Desai, 2002; Gray, 2000).

Today's students also witness and experience rampant privacy violations from online platforms. Contrary to claims that younger generations care less about privacy, evidence indicates that young people in the US as recent as millennials are as concerned about privacy as older generations (Hoofnagle et al., 2010). When asked, younger people who don't use virtual reference services have listed privacy concerns as one reason (Connaway et al., 2011; Mawhinney,

2020). Their concerns may be well-founded; the anonymity afforded by online chat is not absolute.

When students come to a university, they are navigating new issues with privacy daily, which may increase their concern for privacy. Many students are making unfamiliar decisions about giving away their personal information, such as registering for services that may include loans, bills, medical care, and banking (Akhtar & Abbasi, 2019; Givens, 2015; Keizer, 2012). Students will also be greeted with scams such as phishing, social media scams, romance scams, and other attempts at stealing their information (Hanoch & Wood, 2021; Mustofa, 2020; Sorrell & Whitty, 2019; Sutton, 2022). Since libraries are part of the ecosystem of university life, libraries should take steps to ensure the privacy of their students.

Like other online information technologies, some chat platforms automatically gather data about users that they might not otherwise share. For instance, it is not uncommon for chat platforms to gather a user's IP address, which can be connected to "geographic data elements such as city, state, zip code, and possibly even the name of a specific institution or Internet service provider through which the user is accessing the Internet" (Mon et al., 2009). Some libraries may design their chat system to proactively seek personally identifiable information as a pre-condition for beginning the reference transaction (Nolen et al., 2012).

There have also been many studies conducted to analyze chat transcriptions for various purposes. Recent articles used chat transcripts and statistics to determine satisfaction or dissatisfaction with chat services. Logan et al. (2019) conducted a study to identify behaviors that contribute to user dissatisfaction and recommend improvements. In Mavodza (2019), the patterns and types of questions being asked and the number of times they were asked were used to interpret chat transactions for staffing purposes and to "organize live chat guidelines in accordance with patron needs." An earlier study examined face-saving communication (methods of reducing embarrassment or perceived loss of social standing) in chat transcripts by librarians and whether it lowered patron anxiety (Owens, 2013).

It is reasonable to ask whether these factors—which are also seen in other online information technologies familiar to students—are associated with privacy concerns that suppress a library-anxious student's preference for remote help from chat.

Research Question and Objective

To investigate this problem, the authors sought to answer the question: *What relationship exists between students' library anxiety and privacy attitudes in the context of library live chat services?* Because little previous research has explored this question, the authors approach the topic from several angles to establish a foundation for future research. This analysis includes:

1. The authors' experiences with chat as reference librarians
2. An exploration of theoretical perspectives of library anxiety and privacy attitudes
3. A discussion of an empirical investigation connecting theory to real-world data

The authors hope that this chapter can encourage future research into the topic of the privacy behaviors of student users of libraries by providing useful analytical tools.

Library Live Chat at UNC Charlotte

To give context to this study, it will be helpful to describe the university and library where this research took place. The University of North Carolina at Charlotte (UNC Charlotte) is the largest university in the Charlotte region and the third largest of 17 institutions in the UNC System. UNC Charlotte is a research-intensive institution with more than 30,000 undergraduate and graduate students and over 1,000 full- and part-time faculty. The J. Murrey Atkins Library at UNC Charlotte is one of the largest research libraries in North Carolina with a permanent staff of about 90. The Research and Instructional Services (RIS) unit consists of 15 full-time librarians. These librarians are responsible for liaison and collection development duties in addition to staffing the Information and Research Desk, where research services are provided. This desk is also staffed by Access Services employees who primarily check out materials and equipment. In design, Atkins Library follows a common reference model described in Gotschall et al. (2021), in which the front desk at the library entrance, in addition to being the primary point of contact for circulation, is also where users might ask reference questions or schedule meetings with reference librarians.

Atkins Library has been providing a virtual chat reference service since the early 2000s. Virtual reference services have primarily been the responsibility of the library's research librarians, although Atkins Library has used several different staffing models over the years. Atkins Library has also used various chat systems including QuestionPoint, Libraryh3lp, and currently the Springshare platform LibChat.

When beginning a chat in LibChat, users are given the option to include their name or to leave the name field blank to remain more anonymous to the library employee. The employee they are chatting with may or may not be anonymous to the user depending on whether the employee is signed in with their individual account or under a shared department account.

There have been internal conversations about whether displaying librarians' names makes the chat service more personal for the user and whether being more personal is desirable. Chat can give a level of anonymity on both ends that cannot be guaranteed in face-to-face interactions. It is an impersonal space that anonymizes people who may often know each other. If anonymity is the goal, personal connection could be counterproductive. For example, many faculty ask questions directly to their subject librarians, but in the authors' experience, some are hesitant to ask questions because they may feel embarrassed for not knowing information they feel they should know. These faculty have implied that they use chat to ask such questions anonymously. Sometimes, library employees may recognize faculty members on chat either by their question or by their name, if shared. In trying to be friendlier and more personal, should the library employee reveal to these faculty seeking anonymity that they do, in fact, know who they are? Would they prefer to know that the person hearing their question is a friendly face, or would asking a potentially embarrassing question to someone you know feel even more embarrassing? This shows the intricacies of protecting patrons' privacy and balancing the anonymity of library employee and patron interactions, especially for users who have anxiety about their questions.

Theoretical Foundations

In the earliest literature of library anxiety, library anxious students were observed to fear that asking questions of a librarian will reveal personal inadequacy or incompetence, and that this will result in negative judgement from the librarian (Bostick, 1992; Kuhlthau, 1988, 1991; Mellon, 1986). In this chapter, this phenomenon is labeled *perceived interpersonal threat* (PIT). This PIT, which is associated with feelings of shame (McAfee, 2018) and moderated by many situational and social factors (Jan et al., 2020; Jiao & Onwuegbuzie, 1997, 1998, 1999, 2002), inhibits the library anxious student's willingness to seek interpersonal help.

The practical literature of reference librarianship includes strategies to reduce library anxiety by reducing the effect of the PIT by various means, (Carlile, 2007, p. 139; Mavodza, 2019; Onwuegbuzie et al., 2004 pp. 268–274) including live chat (Fagan & Desai, 2002; Gray, 2000). However, the effect that live chat services have on library anxiety is difficult to understand using Mellon's (1986) original framework and to measure using the standard Library Anxiety Scale developed by Bostick (1992) because neither incorporate more modern-day computer technologies that allow for remote assistance (Anwar et al., 2004, p. 280; Jiao & Onwuegbuzie, 2004, p. 139; Katapol, 2012, p. 8; Kwon, 2008, p. 120; Van Kampen, 2004, p. 29). To correct for this, this study instead uses

Erfanmanesh et al.'s (2012) Information Seeking Anxiety Scale (ISAS), which has been used in a number of studies (Aghaei et al., 2017; Erfanmanesh, 2016; Erfanmanesh et al., 2014; Khan et al., 2021; Naveed & Ameen, 2016a, 2016b, 2016c, 2017a, 2017b; Rahimi & Bayat, 2015). The ISAS is best understood as an instrument to measure the phenomenon of information seeking anxiety (ISA) specifically among students in the academic context where the library continues to play a role, but where information seeking may also happen outside the library building (Erfanmanesh, 2012, p. 22).

To understand the privacy component of this study, the authors use the specific lens of *information privacy*, as distinguished from physical or general privacy, because this framework is most relevant when discussing "access to individually identifiable personal information" (Smith et al., 2011, p. 990). *Information privacy concern* (IPC) is the primary construct through which an individual's relationship to information privacy is measured (Malhotra et al., 2004; Smith et al., 1996). Like library anxiety, considerable evidence asserts that IPC is moderated by situational and social factors (Altman, 1977; Kayhan & Davis, 2016; Okazaki et al., 2012; Ozdemir, 2017; Petronio, 2002; Taylor & Altman, 1975; Waldo et al., 2010; Xu et al., 2011).

The measurement scales proposed by Xu et al. (2011) provide the most suitable instruments for measuring how social dynamics between individuals and institutions affect the individual's IPC. These instruments build on *Communication Privacy Management* (CPM) theory, which holds that an individual negotiates their decision to reveal private information using, among other variables, their perception of how much risk is associated with potential privacy breaches and how much control they retain over future information sharing (Petronio, 2002). Potential interpersonal privacy risks may include threats of *stigma*, "the assumption that others might negatively evaluate behaviors or opinions of an individual," or *face*, "situations where our disclosures cause us embarrassment, embarrass others in our group" (Petronio, 2002, p. 70), which overlap heavily with the feelings associated with library anxiety in students.

Hypotheses

If the predictions of communication privacy management theory, library anxiety theory, and information seeking anxiety theory are accurate, one should observe a particular effect occurring between information seeking anxiety and the perceived privacy risk of using live chat. One would expect to see that, like with general library anxiety, higher information seeking anxiety in a student creates a perception that library employees represent an interpersonal threat of embarrassment, judgment, or shame. This perceived interpersonal threat targets fears related to *stigma risks* and *face risks* as described by Petronio (2002). Because live chat involves interpersonal communication with a library employee, these risk perceptions would also carry over to the remote context of live chat. When the student judges the privacy safety of live chat by assessing its privacy risks vs. their privacy control, one would therefore expect them to perceive the use of live chat as being riskier. This hypothesis may be summarized as:

H1. Higher information seeking anxiety affects the students' risk-control assessment regarding live chat privacy by increasing the perceived risk.

The null hypothesis of H1 may be stated as:

$H1_0$. Information seeking anxiety has no effect on the students' risk-control assessment of privacy.

Combining this hypothesis with the full communication privacy management model discussed in Xu et al. (2011) results in the proposed model visualized in Figure 1.

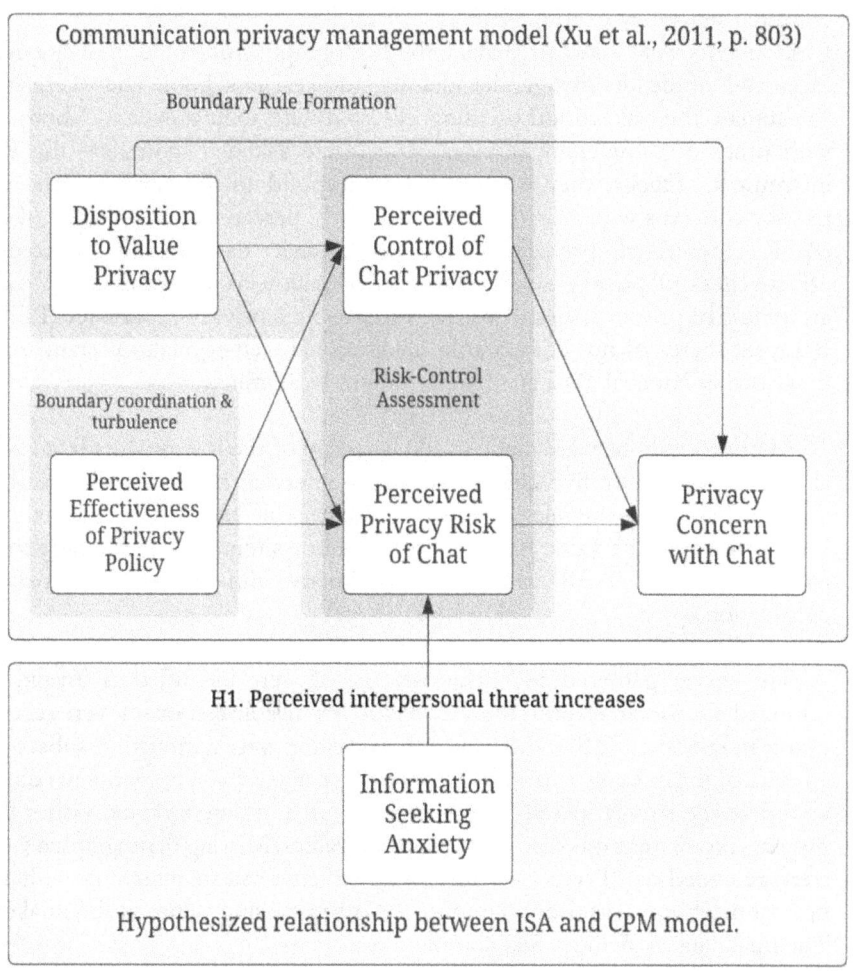

Figure 1: Proposed CPM Model Including ISA

Methods

To test H1 and future hypotheses related to the relationships between ISA, privacy attitudes, and library live chat, the authors constructed a dataset which uses the Information Seeking Anxiety Scale (ISAS) described in Erfanmanesh et al. (2012) and the privacy attitude measurements described in Xu et al. (2011). To build this dataset, the authors distributed an IRB-reviewed survey by email using Qualtrics.com to a stratified random sample of undergraduate and graduate students at or above the age of 18 at UNC Charlotte. Respondents were offered a chance to win one of six $15 gift cards via random drawing as an incentive to participate. Contact information for the random drawing was collected separately from the survey.

Respondents were asked to provide the demographic information of age range, racial and ethnic identity, gender identity, and year in school. They were asked to estimate their likelihood of using live chat with their university library and with other non-university services. They were asked to complete the ISAS instrument. Finally, they were asked to respond to instruments measuring privacy concerns with chat (hereafter: CONC), perceived privacy risks of chat (RISK), perceived privacy control over chat data (CTRL), perceived effectiveness of privacy policy (POLI), disposition to value privacy (VALU), awareness of privacy issues (AWAR), and previous privacy experience (PEXP). Because there is no comparable institutional self-regulation standard as described in Xu et al. (2011), that instrument was omitted.

Minor alterations were made to the language of the instruments from Xu et al. (2011) to specify live chat as the online service being analyzed and the university library as the relevant institution. The scale values were reduced from 7 to 5 to match the values used in ISAS. Minor alterations were made to the language of the ISAS to match vocabulary and phrasing to regional expectations.

The survey gathered 540 responses. Eight were identified as invalid and removed. Due to an error in the survey, too few usable responses were received to one item of the ISAS; responses to this question were removed. A subsequent analysis of the missing values determined that only 0.9% of respondents did not complete the survey in full, and only 0.3% of non-demographic values were missing. No pattern was discovered in these values. Missing demographic values were re-coded as "Prefer not to say." Predictive mean matching with k=5 nearest neighbors was used to impute the other missing values in the final data. The final dataset includes 532 complete responses.

Table 1
Demographic Representation in Sample Compared to University Population

Demographic Category	Sample Frequency	Sample Percent	University Percent
Age Range			
18–24	404	75.9	75.9
25+[a]	128	24.1	23.2
Prefer not to say	1	0.2	–
Race and Ethnicity[b]			
AIAN	10	1.9	0.2
Asian	116	21.8	8.3
Black or African American	60	11.3	16.2
Hispanic or Latino	60	11.3	11.3
NHOPI	2	0.4	0.1
White	339	63.7	51.3
Unknown/PNS	24	4.5	1.8
Gender Identity			
Woman/Female	302	56.8	50.1
Man/Male	173	32.5	49.9
Different Response	16	3.0	–
Prefer not to say	41	7.7	–
Class Status			
Freshman	60	11.3	11.7
Sophomore	92	17.3	16.8
Junior	115	21.6	22.3
Senior	127	23.9	25.8
Graduate Student	130	24.4	20.8
Other	5	0.9	2.2
Prefer not to say	3	0.6	–

Note. Under *Race and Ethnicity*, AIAN abbreviates "American Indian or Alaska Native," NHOPI abbreviates "Native Hawaiian or Other Pacific Islander," and PNS abbreviates "Prefers not to say." University demographics were gathered from the UNC Charlotte Institutional Research Analytics Fact Book at https://ir-analytics.charlotte.edu/fact-book.
[a] Respondents provided more specific values for age range—25–34, 35–49, and 50+—but these values are presented together to facilitate a direct comparison to university figures.
[b] Respondents were allowed to select multiple responses for race and ethnicity. University demographics record race and ethnicity as a single value.

Analysis

Describing the Data

Demographic data describing responses to the student survey is seen in Table 1. Compared to the population of the university from which the sample was collected, White and Asian students were substantially overrepresented in the sample and respondents with the gender identity of Man/Male were substantially underrepresented. Responses for age range and class status were not substantially dissimilar from the university population.

Very few responses were received from the categories of race-American Indian or Alaska Native, race-Native Hawaiian or Other Pacific Islander, and race-Prefer not to say, so these values are included with race-White in the reference group in controls. Age range and class status were found to have high collinearity ($r_s(530)=.51$, $p<.001$); therefore, class status was selected between the two to serve as the control variable used in tests. Ethnicity-Hispanic or Latino was found to have no significant correlation with other variables, so it was determined to be unnecessary to include it as a control variable.

Individual response values to questions of the ISAS and the multiple privacy attitude scales were analyzed to confirm the validity of treating their summed scores as scale measurements. Consistent with best practices (Hayes & Coutes, 2020; Nájera Catalán, 2019; Revelle & Zinbarg, 2009), McDonald's omega (ω) was used to calculate the internal consistency of each measurement scale and inter-item correlations were investigated. No values of ω were discovered to be below 0.734, indicating sufficient internal consistency. Only POLI was found to have an exceedingly high mean inter-item correlation at 0.812 indicating that questions may not be suitably distinct from one another; however, this value was considered acceptable for analysis as an independent variable (IV) in this study. Each scale measurement was normalized to a 5-point scale by calculating the mean value per respondent to facilitate comparison. Descriptive statistics for each measurement across the dataset were calculated. Only POLI was observed to deviate dramatically from a normal distribution, but this was also considered acceptable for analysis as an IV only. These values and the basic descriptive statistics for each scale measurement can be seen in Table 2. To visualize the shapes of these data, a plot of the median, interquartile range, and distribution for each scale measurement in order of lowest to highest median value is provided using violin plots in Figure 2.

Table 2

Reliability Analyses and Descriptive Statistics for Scale Measurements

Scale	N Items	Reliability Measurements		Descriptive Statistics[a]				
		ω^b	Mean Inter-item Correlation	Mean	Median	Std. Deviation	Skewness	Kurtosis
ISAS	46	0.945	0.274	2.56	2.61	0.64	-0.21	-0.18
CONC	4	0.844	0.575	3.04	3.00	0.97	-0.19	-0.55
RISK	4	0.765	0.448	3.12	3.25	0.83	-0.45	-0.10
CTRL	4	0.798	0.498	3.01	3.00	0.87	-0.06	-0.32
AWAR	3	0.743	0.448	3.20	3.16	0.93	-0.15	-0.49
PEXP	3	0.734	0.396	2.70	2.67	0.93	0.21	-0.72
VALU	3	0.797	0.513	3.29	3.33	0.95	-0.25	-0.53
POLI	3	0.929	0.812	4.08	4.00	0.93	-1.43	2.29

[a] To determine these values, all scale measurements were normalized to a five-point scale by summing all responses and dividing by the number of items in the measurement instrument.
[b] McDonald's omega.

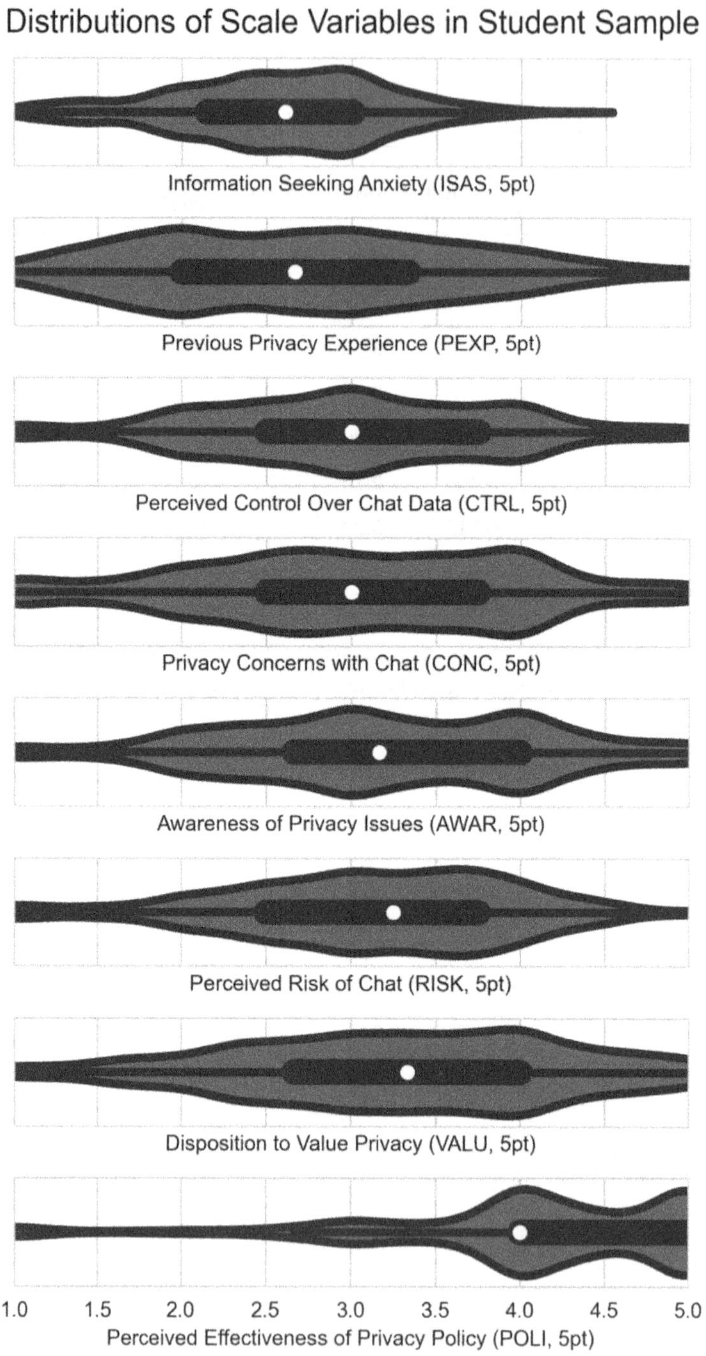
Distributions of Scale Variables in Student Sample
Information Seeking Anxiety (ISAS, 5pt)
Previous Privacy Experience (PEXP, 5pt)
Perceived Control Over Chat Data (CTRL, 5pt)
Privacy Concerns with Chat (CONC, 5pt)
Awareness of Privacy Issues (AWAR, 5pt)
Perceived Risk of Chat (RISK, 5pt)
Disposition to Value Privacy (VALU, 5pt)
1.0
1.5
2.0
2.5
3.0
3.5
4.0
4.5
5.0
Perceived Effectiveness of Privacy Policy (POLI, 5pt)

Figure 2: Distributions of Scale Variables in Student Sample

Testing Hypotheses

Following the CPM model proposed in Xu et al. (2011), both RISK and CTRL should both be affected by POLI and VALU, but RISK and CTRL are determined independently from each other. This study proposes to add Information Seeking Anxiety (ISAS), as a predictor alongside POLI and VALU, with a predicted effect on RISK and a potential additional effect on CTRL. As Xu et al. (2011) suggests, PEXP and AWAR should be used as controls.

To reduce the Type 1 error rate from multiple tests, a multivariate analysis of covariance was performed in which RISK and CTRL were dependent variables. Independent variables were ISAS, VALU, POLI, PEXP, and AWAR, plus several demographic variables found to have some correlation with privacy attitudes. An alpha threshold of .05 was used to determine statistical significance. Results from this analysis can be seen in Table 3.

Table 3

Multivariate Analysis of Covariance for Perceived Privacy Risk and Perceived Control Over Data

Predictor	Multivariate Model					RISK Model[a]			CTRL Model[b]		
	Pillai's Trace	F	df	Error df	Sig.	β	F	Sig.	β	F	Sig.
Intercept	.164	51.186	2	521	<.001	1.783	55.975	<.001	1.698	33.429	<.001
ISAS	.027	7.223	2	521	<.001	.184	14.47	<.001	-.034	.321	.571
VALU	.250	86.680	2	521	<.001	.448	167.4	<.001	.033	.614	.434
POLI	.094	27.035	2	521	<.001	-.145	20.328	<.001	.251	40.455	<.001
AWAR	.030	8.061	2	521	<.001	-.052	1.909	.168	.180	15.391	<.001
PEXP	.012	3.243	2	521	.040	.030	.775	.379	-.105	6.187	.013
race-Asian	.013	3.399	2	521	.034	.161	4.719	.030	.105	1.318	.251
race-Black or AA	.010	2.513	2	521	.082	.174	3.437	.064	.117	1.019	.313
gender-man/male	.009	2.463	2	521	.086	.057	.781	.377	-.171	4.567	.033
class status-graduate student	.002	.541	2	521	.582	-.048	.454	.501	-.062	.489	.485

[a] $R^2 = .354$
[b] $R^2 = .117$

The following variables were found to have a statistically significant effect on the model: AWAR (Pillai's Trace=.030, F2,521=8.061, p<.001); PEXP (Pillai's Trace=.012, F2,521=3.243, p=.04); VALU (Pillai's Trace=.250, $F_{2,521}$=86.680, p<.001); POLI (Pillai's Trace=.094, $F_{2,521}$=27.035, p<.001); ISAS (Pillai's Trace=.027, $F_{2,521}$)=7.223, p<.001); and race-Asian (Pillai's Trace=.013, $F_{2,521}$=3.399, p=.034). The model was discovered to have a significant effect on both RISK ($F_{9,522}$=31.718, p<.001) and CTRL ($F_{9,522}$=5.288, p<.001). However, the explanatory power differed substantially between RISK (R^2=.354) and CTRL (R^2=.117).

Diagnostic tests were performed to ensure that necessary assumptions were met. Box's M value of 32.984 (p=.596) retained the null hypothesis that covariance matrices between groups were assumed to be equal. Levene's F tests against the model retained the null hypothesis that error variance was equal across groups for both RISK ($F1_{2,519}$=.725, p=.727) and CTRL ($F_{12,519}$=.763, p=.689). The standardized residuals of both models were examined via normal Q-Q plots which did not indicate any significant deviation from normality. Kolmogorov-Smirnov normality tests were performed and retained the null hypotheses that residuals were normally distributed for both RISK (D_{532}=.036, p=.092) and CTRL (D_{532}=.025, p=.200). These results indicate that assumptions were met.

Holding all other variables constant, a one-point increase in the five-point ISAS was associated with an increase in the five-point RISK scale of 0.184 points but had no significant association with CTRL. The effect of ISAS on RISK exceeded that of POLI but trailed that of VALU. This result supports hypothesis H1 that information seeking anxiety increases the perceived privacy risk of live chat and rejects $H1_0$.

Discussion

Empirical Findings

Integrating the predictions of communication privacy management theory, library anxiety theory, and information seeking anxiety theory, the authors proposed that the perceived interpersonal threat present in library anxiety would carry to information seeking anxiety and that this sense of threat would affect the privacy risk-control assessment with library live chat by increasing the sense of perceived risk from privacy loss. In other words, students who were more worried about others judging them for having poor research skills would feel like they had more to lose from a lack of privacy and thus be more likely to think that using live chat was dangerous to their privacy.

Our analysis suggests that this hypothesis was indeed observed within the student responses to the survey. As information seeking anxiety increased, so too did the perception of privacy risk, even controlling for factors like other privacy attitudes and demographic differences.

This finding has interesting implications. It suggests that, although chat is assumed to be a preferable option for embarrassed students because of its increased anonymity, the more embarrassed students also felt a higher sense of privacy danger from chat. While one cannot say yet that this has an inhibiting effect on their use of chat, it certainly does suggest the potential. In the communication privacy management model, higher perception of risk is strongly associated with higher privacy concerns with a service in general. It also suggests that other relationships predicted between library anxiety, information seeking anxiety, and communication privacy management might exist.

Implications for Research

This study observed that increased information seeking anxiety is associated with a higher perception of privacy risk with chat. This finding lends credence to two conclusions. First, the use of Information Seeking Anxiety theory is a useful framework for investigating library anxiety in remote contexts. Second, the use of Communication Privacy Management theory is a useful framework for investigating the relationship between library anxiety/information seeking anxiety and privacy attitudes. More research is needed that explores other relationships between these theories. The authors hope to continue to explore these relationships using the dataset created for this study. Specifically, the direct relationship between information seeking anxiety and information privacy concerns should be examined and whether that relationship affects the use of live chat.

Implications for Practice

Library employees with a role involving live chat should familiarize themselves with the privacy concerns of students. These data suggest that interpersonal reference services like chat are sites for a higher perception of privacy risk. Librarians should take extra precautions to ensure perceived privacy risks of chat are proactively addressed. Reference strategies that do not depend on interpersonal interactions may be preferable to students with high information seeking anxiety. Continue to use open and welcoming language in online chats to reduce any anxiety students may have when asking for help, but use caution with language that could indicate to users that their chat could be connected to real-life social networks, as these findings indicate that a perception of interpersonal threat motivates a sense of higher privacy risk. For example, reconsider any practice of asking for identifying information (such as name, class, or instructor) that is not proactively volunteered. Likewise, avoid indicating that you are familiar with potentially identifying information, as you might by saying "I know your instructor well," "I remember you from my instruction session," or "You must be in Dr. Smith's class," etc.). Finally, one way you can reduce the need for information-anxious students to use interpersonal methods to ask for help is by assessing the chats for common questions and proactively updating the library's web pages with that information in places users are most inclined to look.

Novelty

In the context of libraries, privacy is most often discussed through the lenses of information security or ethical practices (Magi & Garner, 2015), but literature which examines the effect of privacy on user attitudes and behavior regarding library services is scant. To the authors' knowledge, this chapter is the first to apply communication privacy management theory to the context of library science and the experience of library users. In doing so, the authors also believe it to be the first time that a theoretical framework has been proposed to understand the behavior of library users in response to privacy attitudes.

This chapter is also novel in the literature of library anxiety, information seeking anxiety, and information privacy concerns. While much work has been done to understand the social antecedents of library anxiety and information seeking anxiety, here the authors show a novel example of a social consequence regarding privacy. Conversely, while the majority of research about information privacy concerns investigate its social consequences, here the authors show a novel example of a social antecedent regarding libraries and information seeking anxiety.

Conclusion

This chapter discusses experiential, theoretical, and empirical connections between library anxiety, privacy attitudes, and library live chat. The authors' own experiences with providing live chat and analyzing chat data suggested that privacy concerns may reduce students' willingness to use the live chat service. The literature of library anxiety and information privacy was explored to find theoretical evidence upon which to build empirical analyses. A specific connection was proposed between the perceived interpersonal threat in library anxiety, the application of information seeking anxiety to remote chat services, and the privacy risk-control assessment of communication privacy management theory. This connection predicted that anxiety related to seeking information leads to a perception that one's personal privacy is at greater risk when using library live chat.

A dataset was constructed to test this prediction. It was revealed that the predicted effect was indeed observed. This suggests that the theoretical frameworks proposed here are a suitable foundation to explore similar relationships between privacy attitudes and the use of library reference services. These data suggest that interpersonal reference services like chat are sites for a higher perception of privacy risk. Implications for future research and library practice are explored.

References

Aghaei, F., Soleymani, M. R., & Rizi, H. A. (2017). Information seeking anxiety among M.A. students of Isfahan University of Medical Sciences. *Journal of Education and Health Promotion, 6*:14. https://doi.org/10.4103/jehp.jehp_88_14

Akhtar, S., & Abbasi, A. (2019). *Privacy: Developmental, clinical, and cultural realms.* Routledge.

Altman, I. (1977). Privacy regulation: Culturally universal or culturally specific? *Journal of Social Issues, 33*(3), 66–84. https://doi.org/10.1111/j.1540-4560.1977.tb01883.x

Anwar, M. A., Al-Kandari, N. M., & Al-Qallaf, C. L. (2004). Use of Bostick's Library Anxiety Scale on undergraduate biological sciences students of Kuwait University. *Library & Information Science Research, 26*(2), 266–283. https://doi.org/10.1016/j.lisr.2004.01.007

Bostick, S. L. (1992). *The development and validation of the Library Anxiety Scale* (Order No. 9310624) [Doctoral Dissertation, Wayne State University] ProQuest Dissertation and Theses Global. https://www.proquest.com/docview/304002616/abstract/D3EC83A3F2D741B4PQ/1

Brown, L. J. (2011). Trending now—reference librarians: How reference librarians work to prevent library anxiety. *Journal of Library Administration, 51*(3), 309–317. https://doi.org/10.1080/01930826.2011.556950

Carlile, H. (2007). The implications of library anxiety for academic reference services: A review of literature. *Australian Academic & Research Libraries, 38*(2), 129–147. https://doi.org/10.1080/00048623.2007.10721282

Connaway, L. S., Radford, M. L., & OCLC Online Computer Library Center. (2011). *Seeking synchronicity: Revelations and recommendations for virtual reference.* OCLC Research. https://doi.org/10.25333/C3QK98

Erfanmanesh, M. (2016). Information seeking anxiety: Effects of gender, level of study and age. *Library Philosophy and Practice, 1317*, 1–20.

Erfanmanesh, M., Abrizah, A., & Karim, N. H. A. (2012). Development and validation of the Information Seeking Anxiety scale. *Malaysian Journal of Library & Information Science, 17*(1), 21–39.

Erfanmanesh, M., Abrizah, A., & Karim, N. H. A. (2014). The prevalence and correlates of information seeking anxiety in postgraduate students. *Malaysian Journal of Library & Information Science, 19*(2), 69–82.

Fagan, J. C., & Desai, C. M. (2002). Communication strategies for instant messaging and chat reference services. *The Reference Librarian, 38*(79–80), 121–155. https://doi.org/10.1300/J120v38n79_09

Givens, C. L. (2015). *Information privacy fundamentals for librarians and information professionals*. Rowman & Littlefield.

Gotschall, T., Gillum, S., Herring, P., Lambert, C., Collins, R., & Dexter, N. (2021). When one library door closes, another virtual one opens: A team response to the remote library, *Medical Reference Services Quarterly, 40*(1), 11–22. https://doi.org/10.1080/02763869.2021.1873612

Gray, S. M. (2000). Virtual reference services: Directions and agendas. *Reference & User Services Quarterly, 39*(4), 365–375. https://www.jstor.org/stable/20863842

Hanoch, Y., & Wood, S. (2021). The scams among us: Who falls prey and why. *Current Directions in Psychological Science: a Journal of the American Psychological Society, 30*(3), 260–266. https://doi.org/10.1177/0963721421995489

Hayes, A. F., & Coutts, J. J. (2020). Use omega rather than Cronbach's alpha for estimating reliability. But…. *Communication Methods and Measures, 14*(1), 1–24. https://doi.org/10.1080/19312458.2020.1718629

Hoofnagle, C. J., King, J., Li, S., & Turow, J. (2010). How different are young adults from older adults when it comes to information privacy attitudes and policies? *Social Science Research Network*, 1–20. https://doi.org/10.2139/ssrn.1589864

Jan, S. U., Anwar, M. A., & Warraich, N. F. (2020). The relationship between emotional intelligence, library anxiety, and academic achievement among the university students. *Journal of Librarianship and Information Science, 52*(1), 237–248. https://doi.org/10.1177/0961000618790629

Jiao, Q. G., & Onwuegbuzie, A. J. (1997). Antecedents of library anxiety. *The Library Quarterly: Information, Community, Policy, 67*(4), 372–389. https://doi.org/10.1086/629972

Jiao, Q. G., & Onwuegbuzie, A. J. (1998). Perfectionism and library anxiety among graduate students. *The Journal of Academic Librarianship, 24*(5), 365–371. https://doi.org/10.1016/S0099-1333(98)90073-8

Jiao, Q. G., & Onwuegbuzie, A. J. (1999). Self-perception and library anxiety: An empirical study. *Library Review, 48*(3), 140–147. https://doi.org/10.1108/00242539910270312

Jiao, Q. G., & Onwuegbuzie, A. J. (2002). Dimensions of library anxiety and social interdependence: Implications for library services. *Library Review*, *51*(2), 71–78. https://doi.org/10.1108/00242530210418837

Jiao, Q. G., & Onwuegbuzie, A. J. (2004). The impact of information technology on library anxiety: The role of computer attitudes. *Information Technology and Libraries*, *23*(4), 138–144. https://doi.org/10.6017/ital.v23i4.9655

Kayhan, V. O., & Davis, C. J. (2016). Situational privacy concerns and antecedent factors. *Journal of Computer Information Systems*, *56*(3), 228–237. https://doi.org/10.1080/08874417.2016.1153913

Keizer, G. (2012). *Privacy* (1st ed.). Picador.

Khan, S., Anwar, M. A., & Naveed, M. A. (2021). Prevalence and levels of information seeking anxiety among business students. *Library Philosophy and Practice*, 1–13.

Kuhlthau, C. C. (1988). Developing a model of the library search process: Cognitive and affective aspects. *RQ*, *28*(2), 232–242. https://www.jstor.org/stable/25828262

Kuhlthau, C. C. (1991). Inside the search process: Information seeking from the user's perspective. *Journal of the American Society for Information Science*, *42*(5), 361–371. https://doi.org/10.1002/(SICI)1097-4571(199106)42:5%3C361::AID-ASI6%3E3.0.CO;2-%23

Kwon, N. (2008). A mixed-methods investigation of the relationship between critical thinking and library anxiety among undergraduate students in their information search process. *College & Research Libraries*, *69*(2) https://doi.org/10.5860/crl.69.2.117

Logan, J., Barrett, K., & Pagotto, S. (2019). Dissatisfaction in chat reference users: A transcript analysis study. *College & Research Libraries*, *80*(7), 925. https://doi.org/10.5860/crl.80.7.925

Magi, T., & Garnar, M. (2015). *A history of ALA policy on intellectual freedom: A supplement to the intellectual freedom manual*. American Library Association.

Malhotra, N. K., Kim, S. S., & Agarwal, J. (2004). Internet users' information privacy concerns (IUIPC): The construct, the scale, and a causal model. *Information Systems Research*, *15*(4), 336–355. https://www.jstor.org/stable/23015787

Mavodza, J. (2019). Interpreting library chat reference service transactions. *The Reference Librarian*, *60*(2), 122–133. https://doi.org/10.1080/02763877.2019.1572571

Mawhinney, T. (2020). User preferences related to virtual reference services in an academic library. *The Journal of Academic Librarianship*, *46*(1), 1–8 https://doi.org/10.1016/j.acalib.2019.102094

McAfee, E. L. (2018). Shame: The emotional basis of library anxiety. *College & Research Libraries*, *79*(2), 237–256. https://doi.org/10.5860/crl.79.2.237

Mellon, C. A. (1986). Library anxiety: A grounded theory and its development. *College & Research Libraries*, *47*(2), 160–165. https://doi.org/10.5860/crl_47_02_160

Mon, L., Bishop, B. W., McClure, C. R., McGilvray, J., Most, L., Milas, T. P., & Snead, J. T. (2009). The geography of virtual questioning. *The Library Quarterly*, *79*(4), 393–420. https://doi.org/10.1086/605381

Mustofa, R. H. (2020). Is big data security essential for students to understand? *Holistica: Journal of Business and Public Administration*, *11*(2), 161–170. https://doi.org/10.2478/hjbpa-2020-0026

Nájera Catalán, H. E. (2019). Reliability, population classification and weighting in multidimensional poverty measurement: A Monte Carlo study. *Social Indicators Research*, *142*(3), 887–910. https://doi.org/10.1007/s11205-018-1950-z

Naveed, M. A., & Ameen, K. (2016a). A mixed-method investigation of information seeking anxiety in Pakistani research students. *Pakistan and Information Science Journal*, *47*(1), 1–10.

Naveed, M. A., & Ameen, K. (2016b). Information seeking anxiety among postgraduate students of university. *Journal of Behavioral Sciences*, *(26)*1, 142–154. https://doi.org/10.4103%2Fjehp.jehp_88_14

Naveed, M. A., & Ameen, K. (2016c). Measuring levels of students' anxiety in information seeking tasks. *Pakistan Journal of International Management & Libraries*, *17*, 56–68. http://escholar.umt.edu.pk:8080/jspui/handle/123456789/3342

Naveed, M. A., & Ameen, K. (2017a). Determining the prevalence and correlates of information seeking anxiety among postgraduates in Pakistan. *Libri*, *67*(3), 205–214. https://doi.org/10.1515/libri-2016-0017

Naveed, M. A., & Ameen, K. (2017b). A cross-cultural evaluation of the psychometric properties of Information Seeking Anxiety Scale in Pakistani environment. *Malaysian Journal of Library & Information Science*, *22*(3), 35–51. https://doi.org/10.22452/mjlis.vol22no3.3

Nolen, D. S., Powers, A. C., Zhang, L., Xu, Y., Cannady, R. E., & Li, J. (2012). Moving beyond assumptions: The use of virtual reference data in an academic library. *Portal: Libraries and the Academy*, *12*(1), 23–40. https://doi.org/10.1353/pla.2012.0006

Okazaki, S., Navarro-Bailón, M. Á., & Molina-Castillo, F.-J. (2012). Privacy concerns in quick response code mobile promotion: The role of social anxiety and situational involvement. *International Journal of Electronic Commerce*, *16*(4), 91–120. https://doi.org/10.2753/JEC1086-4415160404

Onwuegbuzie, A. J., Jiao, Q. G., & Bostick, S. L. (2004). *Library anxiety: Theory, research, and applications*. University Press of America.

Owens, T. M. (2013). Communication, face saving, and anxiety at an academic library's virtual reference service. *Internet Reference Services Quarterly*, *18*(2), 139–168. https://doi.org/10.1080/10875301.2013.809043

Ozdemir, Z. D., Jeff Smith, H., & Benamati, J. H. (2017). Antecedents and outcomes of information privacy concerns in a peer context: An exploratory study. *European Journal of Information Systems*, *26*(6), 642–660. https://doi.org/10.1057/s41303-017-0056-z

Petronio, S. (2002). *Boundaries of privacy: Dialectics of disclosure*. State University of New York Press. https://muse.jhu.edu/book/4588

Rahimi, M., & Bayat, Z. (2015). The relationship between online information seeking anxiety and English reading proficiency across gender. In *Handbook of research on individual differences in computer-assisted language learning* (pp. 449–468). IGI Global.

Revelle, W., & Zinbarg, R. E. (2009). Coefficients alpha, beta, omega, and the glb: Comments on Sijtsma. *Psychometrika*, *74*(1), 145–154. https://doi.org/10.1007/s11336-008-9102-z

Reference and User Services Association (RUSA). (2017). *Guidelines for implementing and maintaining virtual reference services*. RUSA. https://www.ala.org/rusa/sites/ala.org.rusa/files/content/GuidelinesVirtualReference_2017.pdf

Smith, H. J., Dinev, T., & Xu, H. (2011). Information privacy research: An interdisciplinary review. *MIS Quarterly*, 989–1015. https://doi.org/10.2307/41409970

Smith, H. J., Milberg, S. J., & Burke, S. J. (1996). Information privacy: Measuring individuals' concerns about organizational practices. *MIS Quarterly*, *20*(2), 167–196. https://doi.org/10.2307/249477

Sutton, H. (2022). Protect students and staff from scams by providing proactive education, social media outreach. *Student Affairs Today*, *24*(10), 5–11. https://doi.org/10.1002/say.31010

Taylor, D. A., & Altman, I. (1975). Self-disclosure as a function of reward-cost outcomes. *Sociometry*, *38*(1), 18–31. https://doi.org/10.2307/2786231

University of North Carolina at Charlotte. (n.d.). *About UNC Charlotte*. https://admissions.charlotte.edu/about-unc-charlotte

University of North Carolina at Charlotte Institutional Research Analytics. (n.d.) *Fact book: Fall 2020 – present. Institutional Research Analytics*. Institutional Research Analytics. https://ir-analytics.charlotte.edu/factbook

Van Kampen, D. J. (2004). Development and validation of the multidimensional library anxiety scale. *College & Research Libraries, 65*(1), 28–34. https://doi.org/10.5860/crl.65.1.28

Waldo, J., Lin, H. S., & Millett, L. I. (2010). Engaging privacy and information technology in a digital age: Executive summary. *Journal of Privacy and Confidentiality, 2*(1), 1–15. https://doi.org/10.29012/jpc.v2i1.580

Xu, H., Dinev, T., Smith, J., & Hart, P. (2011). Information privacy concerns: Linking individual perceptions with institutional privacy assurances. *Journal of the Association for Information Systems, 12*(12). http://doi.org/10.17705/1jais.00281

Part 3: How Can We Transform Our Pedagogy with Privacy in Mind?

Visible Bruises: Domestic Violence and Trauma-Informed Instruction in Remote Learning Environments

Jennifer Lynn Reichart

The pivot to remote instruction opened doors for the creation and adoption of new best practices in online and distance programming in higher education. However, this pivot also opened doors right into students' private lives and homes. New policies and practices on attendance, participation, grading, camera use, proctoring, and recording have unintentionally created inequities for many underserved students. In one of the most egregious ways, the privacy and safety of students have been compromised for those learning remotely from a place where domestic violence exists in primary or secondary relationships within the household. For domestic abuse survivors, remote instruction not only removed a potential safe haven on campus but also opened the opportunity for fellow students and faculty to directly witness the effects of trauma or even the trauma itself.

Navigating these complex situations requires a solid understanding of Title IX, mandated reporting, and digital privacy; however, a refresher on these policies and laws was rarely a priority during the remote pivot. Many instructors are murky on the requirements and ethics of mandated reporting. Still fewer are familiar with both Title IX law as well as digital transformation fundamentals and digital privacy considerations. Further compounding this issue is that faculty were also not previously equipped to confront both community trauma from COVID-19 and domestic violence trauma among students. Overnight, faculty felt the pressure to become experts in online teaching and proctoring, student data privacy, FERPA, HIPAA, Title IX, and trauma-informed care and instruction. Any misstep can lead to compound trauma or what Freyd and Birrell (2013) have defined as betrayal trauma. Institutional betrayal as compound trauma in higher education can create more inequitable and deleterious outcomes for students learning remotely. This chapter will explore various scenarios of privacy, safety, and confidentiality breaches in remote instruction within homes suffering from domestic violence and how an understanding of trauma-informed instruction, mandated reporting, and institutional betrayal can create more equitable solutions for online students.

Domestic violence is a sensitive subject. Many professionals in education and social services do not know the patterns of violence and the terminology that defines the accompanying trauma. The investigation of this topic relies on a broad-based understanding of the mechanics of domestic violence within the higher education system. The author's own positionality and experiences as an educator have informed and influenced this work, so this chapter begins with this positionality. The inequities experienced by domestic violence survivors are ultimately relationship-based, and to understand these complex relationships, it is necessary to first understand definitions and contexts.

This chapter will first introduce the author's positionality, definitions for understanding, and contexts before relevance. The chapter then employs case study methodology to provide readers from a broad audience (K-12, higher education, community organizations, etc.) with the opportunity to engage with real-world problems in online learning and domestic violence. The chapter concludes with insights into the case studies and recommendations.

Positionality

My positionality and experience as a faculty developer, writer, and researcher have impacted my work in the areas of trauma stewardship and what I call radical faculty self-care. I served many years as a community college developmental instructor, teaching underserved and underprepared students from many backgrounds. I am a compassionate human being, and as my doctoral research was a study of the interstitial relationships between student sexual assault survivors disclosing to faculty as mandated reporters, I expressed the importance of Title IX on my first teaching days of the semester while going over the syllabus with my students. I believe that it is a combination of my own status as a survivor, my compassionate nature as a teacher, my transparency with my students about my research topics, and my emphasis on the importance and protections of Title IX upon first meeting students that led to a high number of student sexual assault and domestic violence disclosures to me. I vividly remember students asking me to come into the hall so that they could lift their shirts and show me their bruises. While the violence took place inside the home, and they wore clothing and makeup to cover the undeniable abuse, these students chose me as a safe person to bear witness to their trauma in the relative safety and reprieve of our physical campus. Furthermore, my colleagues and I have experienced the moral dilemmas of honoring their relationships with their students and the confidentiality of these students who disclose yet request our discretion and the legal obligations of the mandated reporting of sexual violence. Title IX does not address the full murkiness of these relationships, and the consequences of institutional betrayal to students when reporting is not the best option.

These experiences have helped me to become a participatory action researcher and a case study educator on sexual violence in higher education. As a trauma counselor, trauma-informed instructor, and trauma researcher, I subscribe to the survivor-centered approach, doing my best not to misappropriate survivors' narratives while sharing them so that their stories can help practitioners and other survivors move forward. In combining this ethos with case study methodology, I have provided mini case studies in this chapter so that readers may reflect on different scenarios of domestic violence in online classrooms and explore options for how to best assist these students within the contexts of the situation. All cases here are based on real events, use synonymous names, and have been represented with the permission of the survivors.

Definitions

Domestic Violence

Over the years and throughout organizations, the types of abuse experienced by domestic violence survivors have not changed, but the vehicles of abuse have changed drastically. For instance, on-campus stalking still happens, but it now usually involves at least one element of cyberstalking. According to the National Domestic Violence Hotline, "Domestic violence (also referred to as intimate partner violence (IPV), dating abuse, or relationship abuse) is a pattern of behaviors used by one partner to maintain power and control over another partner in an intimate relationship" (n.d.-a). Power and control are the key elements of all types of domestic violence. Domestic violence educators describe several types of abuse including but not limited to physical, sexual, mental, emotional, financial/economic, verbal, and spiritual. Among these, many specific patterns of domestic violence can be found, such as isolation, intimidation, gaslighting, neglect, and stalking. A new type of violence is now recognized: technological abuse, which can involve restricting access to computer files, distributing information and images without consent, and cyberstalking. As we move toward an ever-increasing online teaching and learning environment, it is important to understand the types of abuse that can take place in that environment and how that abuse needs to be addressed, reported, and resolved, whether disclosed by a student or directly witnessed by an instructor.

Primary Trauma

There are multiple, sometimes competing and confusing definitions of trauma. To begin, the term "trauma" can be used interchangeably to refer to both the traumatic event itself and the body of changes in attitude, expression, belief, and value system the traumatic event causes in individuals or groups. The National Institute of Mental Health (NIMH) defines a traumatic event as "a shocking, scary, or dangerous experience that can affect someone emotionally and physically" (n.d.). At the same time, the Substance Abuse and Mental Health Services Administration (SAMHSA) provides some basic definitions for the changes in personhood. SAMHSA describes individual trauma as resulting from "an event, series of events, or set of circumstances that is experienced by an individual as physically or emotionally harmful or life-threatening and that has lasting adverse effects on the individual's functioning and mental, physical, social, emotional, or spiritual well-being" (Delphin-Rittmon, 2022, para 4.). Throughout my career, I have adopted or developed my own definitions of a traumatic event and individual trauma to add to these recognized definitions:

1. a traumatic event is any occurrence in which an individual or group's previous coping mechanisms no longer suffice, causing the individual or group to create a new set of coping mechanisms (these can be positive and/or negative);
2. the body, mind, and spirit's normal reaction to abnormal circumstances.

In any case, primary trauma is the trauma experienced by the direct victim of a traumatic event, such as a survivor of a sexual assault.

Secondary and Vicarious Trauma

However, the trauma is not limited only to students; in being exposed to student trauma, faculty themselves might experience their own trauma. VAWnet, a project of the National Resource Center on Domestic Violence, defines secondary or vicarious trauma as "the emotional effects that can occur when an individual bears witness to the trauma experiences of another. For example, victim advocates may experience secondary traumatic stress from listening empathically to survivors recounting their stories" (n.d.). Usually, secondary trauma results from an intentional disclosure on the part of a student, as in the case of a student voluntarily scheduling time to meet with an instructor to discuss their situation. Vicarious trauma is more of an unintentional witnessing, such as driving to work past a disturbing scene of a car accident. In the case of online and distance education, secondary trauma could result from a student scheduling a private Zoom meeting with an instructor to explain why they are behind on assignments or why they never have their camera on during synchronous sessions. Vicarious trauma could arise from a faculty member directly witnessing domestic violence within the home via audio and/or video capture. This observation could occur synchronously or asynchronously if a video or audio recording is part of a discussion board requirement or the medium of another assignment.

While secondary and vicarious trauma might sound minimal compared to primary trauma in the case of domestic violence, it is far from negligible and can have far-reaching and long-lasting negative effects on the individual bearing witness. VAWnet continues by stating,

> Individuals affected by secondary traumatic stress may themselves experience trauma-related responses as a result of the indirect trauma exposure or may find themselves re-experiencing trauma that they have experienced in their own lives. The cumulative effects of secondary traumatic stress may be seen in both professional and personal life. (2021, para 11.)

These personal reactions, combined with the requirements of mandated reporting for Title IX sexual violence and any abuse of a minor (dual-credit students often opt to take online courses at a local community college or university during high school), can take a huge toll on faculty members who must negotiate what is best for their students and what is required of the law, all in an online space.

Collective Trauma

Across all these definitions, the only constant is change. There are numerous subsets of trauma, including but not limited to compound trauma, complex trauma, and insidious trauma. Most recently, there has been some debate over whether the worldwide COVID-19 pandemic can be considered a collective trauma, with most trauma-informed researchers wholeheartedly including it. Saltzman (2020) defines collective trauma as "an event, or series of events that shatters the experience of safety for a group, or groups, of people" and that "these events are a shared experience that alter the narrative and psyche of a group or community" (para 1). The global pandemic has caused us to reflect on the relative safety of shared spaces on college campuses in terms of contagion; however, in "safely" sending students home, we may have inadvertently sent them back into the home front battlefield. Indeed, many students of all ages have valued the physical educational environment as a safe haven from domestic violence for generations.

Compassion Fatigue

These cumulative effects of vicarious trauma can lead to compassion fatigue, especially where multiple students' disclosures and personal experiences are present. However, vicarious trauma can happen to any instructor who bears witness to student trauma, as we assume that all faculty members are compassionate human beings who care about their students' health, safety, and well-being. VAWnet defines compassion fatigue as a related term used to describe exhaustion and desensitization to violent and traumatic events encountered in professional work or in the media. Both secondary traumatic stress and compassion fatigue can result from bearing witness and connecting empathically to another person's experience and being emotionally present in the face of intense pain. (2021, para 12.)

Institutional Betrayal

Freyd and Birrell (2013) define institutional betrayal as the compound trauma that arises among members of an institution (such as colleges, universities, churches, etc.) as the "institutional failure to prevent sexual assault or to respond supportively when it occurs" (p. 38). When faculty as mandated reporters report disclosed violence to their Title IX office, police officers, and/or the Department of Children and Family Services (DCFS), students can feel betrayed by the faculty they chose to trust when they disclosed to them. Some faculty may see the potential for the erosion or destruction of their teacher-student relationship as a necessary casualty in the battle to protect that student from the violence in their home and/or relationship.

Decolonial Feminist Theory

In evaluating domestic violence relationships of students and the responsibilities of online teaching faculty, the privacy, security, and confidentiality of student data is a correlating concern with the student's health, safety, and well-being. No matter the gender of the students, approaching this equity problem from a critical-feminist theory lens can benefit all. The critical-feminist approach seeks to empower the voices of specific groups which have been historically silenced, what some are now referring to as "Decolonial Feminist Theory" (Manning, 2021). It acknowledges that there is no single truth but that the cumulative knowledge of survivors voicing their experiences can pave the way for newer, more improved policies and practices. It prioritizes women's "freedom, choice, and personal responsibility" over government constraints (UAH, n.d., par. 14). Individualist feminists, also known as "ifeminists,"

> believe that freedom and diversity benefit women, whether or not the choices that particular women make are politically correct. [...] As the cost of freedom, ifeminists accept personal responsibility for their own lives. They do not look to government for privileges any more than they would accept government abuse. Ifeminists want legal equality, and they offer the same respect to men (UAH, n.d., para 14)

This framework of individualist and/or decolonized feminism helps faculty reconsider the needs of students by prioritizing their choices disclosure, action, and safety. It aligns with the domestic violence counseling principle that survivors, especially women, should be trusted to know what is best for them.

Contexts

Vulnerable Populations

While domestic violence does not discriminate across age, gender, race, location, religion, or socioeconomic status, some groups are at a higher risk than others due to environmental and intergenerational factors. Some of the more salient groups of vulnerable populations in the online college environment are discussed here.

Impoverished Students

Many students, unfortunately, struggle with unemployment, food insecurity, and the threat of homelessness. While numerous assertions exist that many individuals are only one paycheck away from being homeless, those familiar with the power and control techniques of domestic violence understand that individuals are often more likely only one relationship away from being homeless. Jones et al. (2012) found in a study that:

> Three themes emerged from the data describing the intersection between respondents' intimate relationships and their situation of homelessness: (1) relationship breakdown; (2) the role and impact of having intimate partners during a period of homelessness; and (3) the nature of the intimate relationship and its impact on housing. The data suggest that aspects of intimate relationships should be considered by social service agencies when addressing a person's situation of homelessness. (p. 101)

A common question asked of domestic violence survivors is why they do not leave an abusive partner and/or why it took them so long to leave. Homelessness is a true threat to domestic violence survivors, as it is "the devil you know." Staying with an abusive partner is often seen as a more viable alternative to living on the street or in a homeless shelter where one can be raped, exposed to HIV/AIDS, or murdered by a stranger or a casual acquaintance within the homeless population.

First-Generation Students

Numerous studies have been conducted into the relative "high-risk" nature of first-generation students in comparison to continuing-generation students in higher education in the United States. The 2017 study conducted by the U.S.

Department of Education's Institute of Education Sciences found that only 20 percent of first-generation college students graduate with a bachelor's degree by the age of 25, compared to 43 percent for continuing-generation students. They also had lower high school grades than their counterparts (Gewertz, 2017). Socioeconomic status, race, gender, and age have all played a part in describing what has been labeled as "non-traditional" students, many of whom are also first-generation. This work takes the environmental and/or community approach to student preparedness, removing the "at-risk" label from students and placing the onus of being underprepared and/or underserved on the educational community. Likewise, students who become victims of sexual assault will not be described as engaging in "risky behavior" but rather being exposed to risk factors within the educational environment.

Unfortunately, the global pandemic unsettled the foundations for first-generation students in many ways, including financial, environmental, and safety situations. Soria et al. (2020) found that:

> First-generation students were more likely than continuing-generation students to experience financial hardships during the pandemic, including lost wages from family members, lost wages from on- or off-campus employment, and increased living and technology expenses. Compared to continuing-generation students, first-generation students are nearly twice as likely to be concerned about paying for their education in fall 2020. Furthermore, first-generation students were less likely to live in safe environments free from abuse (physical, emotional, drug, or alcohol) and more likely to experience food and housing insecurity. First-generation students also experienced higher rates of mental health disorders compared to their peers. (p. 1)

First-generation students, like undocumented students, experience real concerns about their abilities to pay for college while battling other barriers to education and dealing with home life.

Undocumented Students

The threats of domestic violence, intergenerational violence, and homelessness can be more realistic for some groups than others. While the cruelty of divided families was visible via the media during the Trump Administration's removal and relocation of undocumented Latinx immigrants, the threats within these destroyed or threatened households were more concealed. In her recent book, *Uncolonized Latinas: Transforming Our Mindsets and Rising Together,* Valeria Aloe (2022) states,

> Our usual barrier to our healing is we have learned to suppress what we feel. As immigrants and daughters of immigrants, we coped with

trauma, isolation, domestic violence, alcoholism, and more, but we have pushed through in isolation [...] We all carry trauma and pretending our pain is not there will not make it go away. Sooner or later, it comes to the surface..." (p. 217).

Undocumented students, such as those mentioned in the above example of intergenerational Latinx violence, have been at even higher risk during the pandemic due to isolation during sheltering-in-place protocols and the recent Trump administration's policies on immigrants, immigration, and deportation. As stated above, domestic violence centers around more power and control. These extra layers of policy and protocol gave abusers more abusive power and coercive control over their partners and have been a terrifying reality for many Latinas and/or undocumented partners. Immigration status abuse and threat is now a recognized type of domestic violence used to intimidate undocumented partners and keep them quiet, under control, and attached to their (often documented or legal citizen) abusive partner.

Given the number of salient factors affecting online students pre- and post-COVID, teaching faculty must understand multiple types of trauma involving privacy, security, health, and safety in the home, family, community, and institution. Implementing trauma-informed instruction for online students is more necessary than ever before.

Case Study Methodology

While domestic violence is a historical problem, technology is a new one. Due to the changing landscape of higher education, this chapter takes a case study approach to engage the highly interpretive and complicated nature of domestic violence in online education. After reading the two mini case studies, readers should evaluate how they might best help these students by utilizing the given strategies and considerations. These case studies can be examined individually but are best utilized in the traditional case study methodology of employing groups. It is recommended that the case studies be used in a department meeting or professional development session in which faculty from the same disciplines engage with one another on the topic of digital privacy and domestic violence. A facilitator can assign the case studies in the session as they are short, break faculty into pairs or small groups, have them discuss the case studies and debate potential solutions, and then come back to discuss in the larger group. This activity is often referred to as a "Think-Pair-Share." This activity is great for these types of meetings and provides faculty with a model to use in class activities. Participants can engage in person in small groups or online via breakout rooms. Brainstorming solutions first in small groups and then in the larger group before reading the strategies section is recommended. This will fuel solution generation via divergent, creative thinking and encourage dialogue on competing needs and potential solutions. Readers should factor in the unique views of the students, the faculty, and the institutions, the requirements of mandated reporting and Title IX policies, and what is best for the student.

Mini Case Study #1

Miss Phoebe is teaching an asynchronous online course in Communications. As part of the course, students are required to introduce themselves during the first week by uploading a video of themselves talking about their career goals. Victoria, a Latina in her early 20s, is enrolled in the course from the beginning, showing active in the LMS analytics, but does not upload a video and receives a zero for this assignment. For the rest of the semester, Victoria is an active, engaged, and enthusiastic learner, participating in all the asynchronous discussions, assessments, and assignments, including ones requiring videos of herself talking. At the end of the semester, Victoria is sitting between a B+ and A- grade, the zero from the first video introduction assignment being the only points bringing her cumulative course grade down. Miss Phoebe receives an email from Victoria stating that she truly enjoyed the course and did not complete the first assignment because she lives with her boyfriend. During that first week of class, she had visible bruises on her face and neck because her boyfriend had become angry with her for spending money on college tuition and "letting uninvited guests into [his] house" via the online class. Victoria asks if she can make up the assignment. She tells Miss Phoebe that her boyfriend has been very sweet and supportive since he made clear that he would not help her financially with her future tuition. Victoria shares that her boyfriend is much calmer now that he understands that the asynchronous nature of the course meant that others would not be able to see or hear what was happening live in his house. Victoria hints that she is an undocumented student while her boyfriend is a legal citizen, older than her, and white.

Mini Case Study #2

Mr. Burton teaches an online, synchronous course in Economics. He prides himself on his engaging micro-lectures, breakout room discussions, and group activities. Jasmine, a white student in her late 30s, is very talkative and gives great examples of how she connects the course material to her experiences as a lesbian woman. However, she only engages using the chat or audio features in the LMS; she has never once turned on her camera. Mr. Burton's syllabus states that it is the expectation that students will "be actively engaged with their cameras on" during their synchronous online sessions. Still, after working with Jasmine, he realizes that he never wrote a consequence for a student not turning on their camera. He also realizes that "expectation" is not a "requirement." He chooses to let it slide until multiple students private message him to complain about this in one session. Feeling the pressure, Mr. Burton privately messages Jasmine, "Is there a reason you never have your camera on? Please turn your camera on." Instead of turning her camera on, Jasmine immediately and completely leaves the online learning environment. The next day, Mr. Burton receives an email from Jasmine stating, "I was triggered when you called me out yesterday. The reason I don't like to turn my camera on is that I am uncomfortable being on a webcam because I was forced to do pornography as a child by the men in my family. I know this is something I need to work on because everything is online these days, but I'm just not there yet, and you pressuring me that way reminded me of being pressured by other older men in my life. I'm not sure what to do now moving forward in this class." Mr. Burton does not know how to do so either.

The Need for New Strategies

The following section is provided as further considerations within the contexts of the mini case studies. After reviewing the cases, readers should continue to this next section and then return to brainstorming potential solutions on their own or in groups. Faculty are recommended to incorporate these practices into their teaching wheelhouse and prepare for these situations when teaching online courses.

Trauma-Informed Instruction

At a base level, safety is a survival instinct; there is not much that one can accomplish without feeling safe. Minahan (2019) reminds readers:

> Students can't learn unless they feel safe. When it comes to student trauma, there is much that is beyond educators' power, but there is also a great deal they can do to build a supportive and sensitive environment where students feel safe, comfortable, take risks, learn, and even heal.

Trauma-informed instruction is simply that—being supportive of our students' needs knowing that we cannot control or prevent everything. There is a certain amount of letting go to be done in the service of trauma-informed instruction. Just as trauma-informed care is utilized by frontline domestic violence advocates, counselors, and therapists, the wise professional knows that while they might not be able to "check it at the door," there is a professional line that one cannot cross. The even wiser professional and teacher invests more in their own self-care than in their students to best provide the service and support necessary for their most underprepared and underserved students.

Online Safety Planning

One of the most common questions asked of those in abusive relationships is why the abused do not leave the abusive. Fortunately, as practitioners and researchers have learned more, we have a troubling yet stark answer to this question. The National Domestic Violence Hotline states,

> When a survivor leaves their abusive relationship, they threaten the power and control their partner has established over the survivor's agency, which may cause the partner to retaliate in harmful ways. As a result, leaving is often the most dangerous time for survivors of abuse. (n.d.-b)

Again, the decolonial feminist lens asks us to let women make their own decisions, so rather than telling survivors to leave their abusive partners, give them information and let them make their own decision. We should remind ourselves that they know their partners best.

Creating a safety plan can truly be a lifesaver for domestic violence survivors. Safety plans are based upon different categories of "If… Then…" statements involving domestic violence. Depending on one's situation and their partner, a safety plan might mean leaving or staying with the abusive partner; unfortunately, sometimes staying is the safer choice. If a student discloses to an instructor in a synchronous online setting, the burden of proof has been met. This is the time to ask a student about online safety planning. Technological abuse can easily extend into an online student's educational life. While the instructor does not need to help the student create an online safety plan, they can encourage the student to do so. LoveIsRespect.org has an excellent guide called "Who's Spying on Your Computer? Spyware, Surveillance, and Safety for Survivors." This guide is referenced in this chapter, but it should only be shared with a student survivor once they have made it clear that it would be safe to do so. Student survivors can also safely access the National Domestic Violence Hotline online; the site builds safety by allowing website visitors to immediately hit the red X in the top right corner to safely leave the site without leaving it in the browser's history. Once again, it is only safe to share this with a student if there will not be a record of the website link in an email.

Intersectionality Matters

All students are individuals. They have different needs. In reflecting on those needs, cultural considerations need to be made. As in the given case studies, these students' histories, genetics, finances, and more all play a part in how they live in and approach the world. In trying to support all students, instructors should educate themselves on the possible needs of students without drawing particular attention to them while trying to support the student. They should avoid making assumptions or asking questions about students' race, finances, sexual orientation, religion, and age. However, understanding these might help them better interact and connect with students. For example, in many cultures and religions, divorce is not an option, so safety planning might involve minimizing the effects of the abuse instead of leaving it altogether. In any case, safety planning is best left to professional domestic violence counselors who have training in conducting lethality assessments. Compile a list of local and online resources to share with diverse students to reach out to these professionals. Many local and state agencies provide assistance for specific groups such as disabled women who are survivors of domestic violence, and the National Domestic Violence Hotline provides interpretation services in over two hundred languages. As we serve a diverse body of students, some of whom may now be able to take college courses via online instruction for the first time, we need to be cognizant of their needs.

Making Allowances

Sometimes, it may be prudent to give a student an alternative way to complete an assignment or to extend a deadline. This can obviously arise from any number of situations beyond domestic violence. However, having the compassion to help students out in these ways will not only help the students to be more successful in college but can also help them to trust their instructors more. This can make all the difference when they need help. It should be remembered that instructors should only engage with students on this topic if it does not put the student at more risk. One way to circumvent this issue is to make resources available to all students. For instance, resources for domestic violence can be placed in the syllabus or in the LMS along with other key resources on disability services and Title IX policies. This way, the information is accessible to all students at any time, is contained among other information which helps it be more innocuous, and does not single any one student out through email or messaging.

Allowing Choice

Mandated reporting leaves no wiggle room. However, if reporting a domestic violence situation puts a student survivor more at risk, is this truly ethical? This is a question each instructor will have to grapple with themselves if they ever meet with this type of situation. Having said that, decolonial feminism prioritizes choice over governance. If subscribing to this paradigm, instructors can follow this practice if a student discloses to them: explain the nature and requirements of mandated reporting so that students know for the future, give the student the choice of whether they want the instructor to report it, follow through on the student's wishes, and refrain from pressuring the student to take any actions. One well-timed check-in with the student throughout the semester can be taken as a sign of compassion; constant checking in can feel like pressure or manipulation, which the survivor is already experiencing from their partner. Share resources for domestic violence when safe; then leave the student to choose how they will proceed.

Conclusion

The recent pandemic has given us new ways to connect with one another in the online teaching and learning environment and a better understanding of the effects of isolation and trauma on individuals. As we move forward, we can build upon these principles to better understand the unique challenges our students face and assist those who might be in a domestic violence situation.

References

Aloe, V. (2022). *Uncolonized Latinas: Transforming our mindsets and rising together.* New Degree Press.

Delphin-Rittmon, M. E. (2022, June 16). Address trauma and mass violence. Substance Abuse and Mental Health Services Administration. https://www.samhsa.gov/blog/addressing-trauma-mass-violence

Egbert, S., & Camp, S. (2021). Teaching trauma-burdened students: Life-balancing self-care strategies for educators. *Best of TPC 2021 Report.* Magna Publications. https://www.magnapubs.com/wp-content/uploads/2021/10/free-report-Best-of-TPC-2021-opt.pdf

Freyd, J. J. & Birrell, P. J. (2013). *Blind to betrayal: Why we fool ourselves we aren't being fooled.* John Wiley & Sons, Inc.

Gewertz, C. (2017, September 26). First-generation college students face special risks, study finds. *Education Week.* https://www.edweek.org/teaching-learning/first-generation-college-students-face-special-risks-study-finds/2017/09#:~:text=The%20study%20finds%20that%20while,likely%20to%20earn%20bachelor's%20degrees.

Jones, M., Shier, M. L., & Graham, J. R. (2012). Intimate relationships as routes into and out of homelessness: Insights from a Canadian city. *Journal of Social Policy, 41*(1), 101-117. https://doi.org/10.1017/S0047279411000572

LoveIsRespect.org. (n.d.). *Who's spying on your computer? Spyware, surveillance, and safety for survivors.* https://www.loveisrespect.org/wp-content/uploads/2019/11/LIR-Who_Spying-1.pdf

Manning, J. (2021). Decolonial feminist theory: Embracing the gendered colonial difference in management and organisation studies. *Gender, Work & Organization, 28*(4), 1203-1219. https://doi.org/10.1111/gwao.12673

Minahan, J. (2019, October 1). Trauma-informed teaching strategies. *ASCD.* https://www.ascd.org/el/articles/trauma-informed-teaching-strategies

National Domestic Violence Hotline. (n.d.-a). Understand relationship abuse: We're all affected by the issue of domestic violence. https://www.thehotline.org/identify-abuse/understand-relationship-abuse/

National Domestic Violence Hotline. (n.d.-b). Why people stay: It's not as easy as simply walking away. https://www.thehotline.org/support-others/why-people-stay-in-an-abusive-relationship/

National Institute of Mental Health. (n.d.). Coping with traumatic events. https://www.nimh.nih.gov/health/topics/coping-with-traumatic-events

Redford, J., Hoyer, K. M. (2017). *First-generation and continuing-generation college students: A comparison of high school and postsecondary experiences.* Stats in Brief. NCES 2018-009. National Center for Education Statistics.

Saltzman, L. (2020, January 17). Understanding collective trauma: The first step toward healing. *Tulane University School of Social Work.* https://socialwork.tulane.edu/blog/collective-trauma/

Soria, K. M., Horgos, B., Chirikov, I., & Jones-White, D. (2020). *First-generation students' experiences during the COVID-19 pandemic.* Student Experience in the Research University (SERU) Consortium. University of Minnesota Digital Conservancy. https://hdl.handle.net/11299/214934

Substance Abuse and Mental Health Services Administration (SAMHSA). (n.d.). Trauma and violence. https://www.samhsa.gov/trauma-violence

University of Alabama at Huntsville (UAH). (n.d.). *Kinds of feminism.* https://www.uah.edu/woolf/feminism_kinds.htm

VAWnet. (2021). Definitions. *Gender based violence resource library – VAWnet.org.* Retrieved September 19, 2022, from https://vawnet.org/sc/definitions

Pedagogy of Privacy: Inclusive Teaching and Disclosures of Disability

Sarah Whitwell and Samantha Clarke

Colleges and universities have long offered online degree programs, courses, and training opportunities. The ongoing COVID-19 pandemic, however, has made online learning a necessity rather than just an option. In many cases, the rapid shift from in-person to online learning has resulted in unintended breaches of privacy. Accessibility concerns and inequities in student accommodation procedures, for example, force students to disclose disabilities or trauma to navigate online classes successfully. Students experience a variety of invisible circumstances that negatively impact their learning, such as attention or comprehension problems, lived trauma, low vision, impaired hearing, and more.

While postsecondary institutions have services available to support students who have disabilities or who have experienced trauma, access is dependent on students disclosing personal information to secure support. These processes can often be time-intensive, and some students have negative encounters with instructors when they seek institutionally mandated accommodations for learning. While students without disabilities may never need to discuss their personal health or lives with university personnel, students with disabilities or who have experienced trauma report additional time and energy spent connecting with student services on campus, meeting with instructors to discuss accommodations and disclosing information which may be deeply personal or traumatic (Wilks, 2022).

This chapter explores the challenges of invisible barriers in the online classroom and how to leverage inclusive pedagogy to proactively mitigate those barriers, reducing the need for personal disclosures. Inclusive pedagogy prioritizes the creation of a supportive learning environment where all students have equal access to learning. Studies show that students of all backgrounds perform better in an inclusive environment (Coughlan et al., 2019; Hand et al., 2012). There are many ways to cultivate inclusive learning environments. For example, Universal Design for Learning (UDL) offers a framework to eliminate barriers for students by designing learning experiences that are accessible to as many learners as possible. Trauma-informed teaching is a pedagogical practice that recognizes trauma and its impact on the individual, and endeavours to create inclusive learning environments. Finally, Culturally Responsive Pedagogy (CRP) bridges the gap between instructor and student by connecting students' cultures, languages, and lived experiences with what they learn in the classroom; it focuses on what students can do rather than what they cannot. No single method or strategy will make a learning experience accessible to all learners. However, UDL, trauma-informed teaching, and CRP emphasize using diverse teaching methods and having flexibility built into the course.

Literature Review

Research shows that designing online courses with accessibility in mind is beneficial for instructors as well as students, as it can reduce the amount of time spent developing alternate formats and structures to accommodate individual students (Basham et. al., 2010; Cook & Rao, 2018; Johnson-Harris & Mundschenk, 2014; Michael & Trezek, 2006; Taylor, 2016). Literature on Universal Design for Learning highlights the model's benefits for accessibility by increasing student agency to select content, formats, and expressions of learning that best suit their skillsets. However, the ramifications that accommodation procedures have on student privacy are still largely unacknowledged.

Similarly, the implications of trauma-informed teaching and CRP for students with disabilities' experience in the classroom are understudied. Frequently, researchers discuss underdiagnosis of disability due to misunderstandings of students' culture, especially if they have moved from another country and have English as an additional language (Blanks & Smith, 2009; Gallagher, et al., 2011; Scott, et al., 2014). Nonetheless, very few researchers discuss disability as a form of culture or a lived experience—the primary example is a blog post and not a scholarly journal article (Dufour, n.d.).

We argue that the nexus of UDL, trauma-informed teaching, and CRP can mitigate students' privacy concerns by reducing the need for disclosures and building inclusive classrooms where disability is embraced as a key component of society. Bringing together research on accessibility, culture, and trauma can provide deeper insights into the lived experiences of students in the classroom, both in-person and online, to support them holistically. Throughout this chapter, we draw upon our backgrounds as Educational Developers and educators at McMaster University to explore how UDL, trauma-informed teaching, and CRP can eliminate or drastically reduce the need for disclosures of disability and address unintended breaches of privacy in remote learning environments.

Student Accessibility Services at Canadian Institutions

Students in Canada have access to a wide range of options for postsecondary education. Laws supporting the rights of students with disabilities to access postsecondary education have facilitated a significant increase in the number of individuals who enroll in universities, colleges, and other postsecondary institutions. The dramatic upswing of online course offerings that resulted from the COVID-19 pandemic further created opportunities for disabled students who could now access higher education from home. In Ontario alone, there are currently 828 online programs offered by postsecondary institutions (eCampusOntario, 2022).

Postsecondary institutions operate independently and are free to determine their own academic and admissions policies, programs, and staff appointments. However, they are governed by the *Charter of Rights and Freedoms,* as well as provincial human rights statutes regarding the accommodation of students with disabilities. All publicly funded postsecondary institutions in Ontario, for example, must have centres or offices for students with disabilities. These centres or offices are responsible for coordinating services and supports for students with disabilities (Ontario Human Rights Commission, 2002). Although variously titled at different institutions, we will refer to these centres or offices as Student Accessibility Services (SAS) throughout.

The Federal Disability Report (2010) drafted by Human Resources and Skill Development Canada indicated that approximately 15% of university students and 16% of college students identify as having a learning disability. These statistics, however, are incomplete; they do not include individuals who have undiagnosed and/or undisclosed disabilities. Lack of access to healthcare, limited transportation options, or communication barriers are just a few reasons why an individual may have an undiagnosed disability. Stigma, prejudice, and stereotypes may also cause people with a diagnosed disability to avoid making a disclosure to protect their privacy. Beyond incomplete statistics, the real problem is that those who have undiagnosed and/or undisclosed disabilities are effectively cut off from accessing accommodations in postsecondary institutions.

Although the implementation of academic accommodations may vary across institutions, the onus of accessing those accommodations consistently falls to the student. At a bare minimum, the student must disclose their disability to Student Accessibility Services. However, most institutions have a policy that requires formal documentation signed by a registered and regulated health professional (e.g., medical doctor, registered psychologist, registered occupational therapist, registered speech and language pathologist) or a recognized and credible expert (e.g., an institutionally appointed sexual assault response coordinator) to access accommodations (McMaster University, 2020; University of Guelph, 2016; University of Saskatchewan, 2021; Western University, 2019). For many individuals, this is a daunting and time-intensive process (McKenzie, 2015).

Although there may be variations across institutions, most postsecondary institutions follow a similar process. To secure academic accommodations, a student must first notify the institution of their need for accommodations by registering with Student Accessibility Services. The student then completes intake forms, during which they are asked to provide documentation regarding their disability. At McMaster University, the focus is on the functional limitations related to a disability that restrict performance in a postsecondary environment. Officially, students are not required to reveal medical information, though intake forms suggest that "this information can be helpful in completing a thorough assessment for accommodation and support needs" (McMaster University – SAS, n.d.). Even without sharing a diagnosis, a regulated health professional must sign the intake forms, confirming that the student does indeed have a disability. This process can be invasive and intimidating for students, especially those who struggled to secure a diagnosis and may feel that their disability is in question. The process of seeking accommodations then can become a violation of privacy.

Once a student has completed the necessary forms and provided documentation acknowledging their disability, the student then meets with a coordinator from Student Accessibility Services to negotiate appropriate accommodations. Students are frequently involved in this process and are active participants in determining appropriate accommodations. However, final approval does rest with Student Accessibility Services to determine what accommodations will best support the student. This is based on consideration of a student's experienced difficulties and history using accommodations, information from medical documentation, and information regarding course requirements. At the University of Saskatchewan, for example, the "Duty to Accommodate" states that "students must actively participate in developing and implementing strategies related to their own academic success and be open to trying solutions proposed by [SAS]" (University of Saskatchewan, 2021, p. 3).

Already, the process of securing accommodations may seem daunting. Students must gather documentation and meet with Student Accessibility Services to develop an accommodation plan. This can take time away from students' coursework and other obligations. Once an accommodation plan is in place, the student must still go through the process of ensuring those accommodations are implemented in each of their courses. At McMaster University, students use a self-registration portal to activate their accommodation plans for each individual course. The instructor then receives a letter outlining the accommodations granted, which could include consideration for extensions, additional time on tests and exams, recordings of lectures, or leniency for missed classes. Students are responsible for following up with the instructor as needed to ensure that their accommodations are being implemented (McKenzie, 2015).

Accommodations are intended to be strictly confidential and based on functional limitations; instructors are never informed of diagnoses, and they are not supposed to ask. Confidentiality is always a key phrase linked to accommodation policy in order to protect the privacy of students, yet the process of securing accommodations is inherently predicated on disclosure. Students must reveal that they have a disability to activate their accommodations, even if they do not need to share the formal diagnosis. Moreover, students are often expected to negotiate with instructors to ensure that their needs are being met, a burden that students without disabilities do not experience. In the case of extensions, a student may need to inform an instructor for each assignment that they have encountered a barrier and need to activate their accommodations. Even when the accommodation process is functioning as intended, students are forced to share personal information to receive equitable opportunities in the classroom.

There are, of course, scenarios in which the accommodation process does not function as intended. Some instructors falsely believe that accommodations reduce academic rigour and give some students an unfair advantage in the classroom. These instructors can be belligerent, making demands of students with disabilities that are unfair and unwarranted (Olney & Brockleman, 2003). Moving away from an accommodation model to an accessibility model, however, allows all students the opportunity to succeed without burdening students and placing them in a position where they must advocate for equitable treatment.

In talking about moving towards an accessibility model as a way to mitigate privacy concerns, we must first define some key frameworks and pedagogical practices: Universal Design for Learning (UDL), trauma-informed teaching, and Culturally Response Pedagogy (CRP).

Universal Design for Learning

Universal Design involves designing products, buildings, and environments so that they can be accessed and readily used by a variety of users. The idea is to remove barriers through the initial designs by considering diverse needs, rather than overcoming barriers later through individual accommodations or adaptions. Essentially, universal design means creating something with everyone in mind (Rose et al., 2006).

In recent decades, Universal Design has been applied to higher education as Universal Design for Learning (UDL). The Centre for Applied Special Technology (CAST) created the Universal Design for Learning framework and guidelines to help instructors transmit information, and support and foster the growth of knowledge and skills (CAST, 2022). This is accomplished by embedding accessible pedagogy through multiple means of representing information, multiple means for expression of knowledge, and multiple means of engagement in learning (CAST, 2018). Universal Design for Learning recognizes that students are individuals with unique experiences, and that they may have differences in the way they perceive and comprehend information. This is especially important for students with disabilities who may find some forms of representation, expression, and engagement completely inaccessible.

First, "multiple means of representing information" captures the importance of presenting information in a multitude of ways because there is no one way of representing information that will address the needs of all students (Rose et al., 2006). Students with vision impairment, for example, may struggle to access information that is presented only in a visual format. Instructors might consider providing audio files or braille versions of texts. However, physical disabilities are only one consideration, and instructors should also consider students who may be English Language Learners (ELLs), or who come from a cultural background with different classroom experiences. Presenting information in a multitude of ways make it possible for students to engage more fully in the classroom without the need for accommodations.

Second, "multiple means for expression of knowledge" acknowledges that students navigate learning environments and express their learning in different ways. A student with attention deficit disorder (ADD) or attention deficit hyperactivity disorder (ADHD), for example, may have a wide variety of skills but lack the executive functions necessary to achieve long-term goals. There is also the reality that some students express themselves best in one medium over another (Rose et al., 2006). Instructors might consider allowing students to choose from different types of assessments, such as essays, presentations, or multimedia assignments. If instructors provide students with choice in the ways that they demonstrate their learning, they can better support students.

Finally, "multiple means of engagement in learning" recognizes that students have different motivations for learning. Individual variation can be the result of neurology, cultural background, personal experience, and background knowledge. Where one student might be engaged by spontaneity, another may be disengaged or even frightened (Rose et al., 2006). One student may prefer to work independently, and another may enjoy collaborating within a group. As CAST articulates, there is no one means of engagement that will optimally engage every student (CAST, 2018). By providing options, students are given a chance learn on their terms.

Trauma-Informed Teaching

Trauma-informed teaching embraces many of the same strategies as UDL. Trauma-informed teaching recognizes that students have different lived experiences and encourages instructors to proactively consider how trauma may impact learning. The dynamics of complex trauma can negatively impact several executive functions, including inhibitory control (the capacity to regulate strong emotional or impulsive behavioural response), cognitive flexibility (the ability to think about multiple ideas or switch quickly between ideas), and working memory (the ability to process and remember new information). A student who has experienced trauma may struggle with these executive functions and, as a result, have difficulty fully engaging with course content (Barr, 2018). Research indicates that as many as 68% of children experience at least some kind of trauma event, and while many will not experience post-traumatic effects from these experiences, others will carry this trauma forward into adulthood (Cavanaugh, 2016). Understanding how trauma can hinder learning allows instructors to better support students by meeting their individual needs and allowing them to engage with course content in ways that do not cause further trauma.

Trauma-informed teaching is rooted in the understanding that trauma is individual, and a traumatic event for one person may not prompt a trauma response for another. Instructors therefore should consider what content in their course may be triggering for students and provide students with information so that they can make informed decisions about their own well-being while still engaging in learning (Sitler, 2009). For example, an instructor may include content notice in advance of teaching a topic that could be traumatic for some students and articulate that students may choose to opt out of those discussions. Topics like racism, sexual violence, and domestic abuse can prompt a trauma response for certain individuals, and by giving them notice of the topic and providing them with options for how to engage or not, the student is not put in a situation where they must prioritize learning over their mental or physical health.

Culturally Responsive Pedagogy

Culturally Responsive Pedagogy (CRP), as the final framework discussed here, considers the cultural identities and lived experiences of students. It is a direct response to growing concerns over academic achievement differentials on the basis of race, socioeconomic class, and level of English-language ability. Research indicates that racialized students, students from lower socioeconomic classes, and English language learners have long been undervalued in higher education, and their cultural differences are seen as barriers to learning (Vavrus, 2008).

The term CRP was coined by Geneva Gay who recognized the value of aligning academic knowledge and skills with the lived experiences and frames of reference for students. This creates more meaningful learning, and students are more likely to become more engaged (Gay, 2000). In this context, culture refers to the customs, languages, values, beliefs, and achievements of a group of people. Students are inherently shaped by their culture, and it impacts how they make sense of the world and navigate learning environments.

There are five components of culturally responsive teaching. First, instructors should develop knowledge of cultural diversity; they must understand the cultural values and traditions of different racial and ethnic groups and incorporate these into their instruction. Next, instructors should ensure that course content includes a diversity of perspectives. This might mean showcasing readings by individuals of varying race, class, ethnicity, and gender. Students will be better able to see themselves in the curriculum and begin to understand their place in the learning that is taking place. Third, instructors should have the same expectations for all students. All students should be expected to perform at the highest level regardless of race, gender, class, and ethnicity. Fourth, instructors should appreciate different communications styles. Indigenous cultures, for instance, place high value on storytelling and oral history, yet these communication styles have long been derided and considered inferior to other communication styles or record keeping. By embracing these different communication styles, instructors can create more space for students of different backgrounds to participate in the classroom. Finally, instructors should connect course content to students' prior knowledge and cultural experience; there is value in the unique experiences of individuals, and instructors can highlight this through CRP (Gay, 2002).

Strategies for Inclusive Teaching

Combined, UDL, trauma-informed teaching, and CRP, can improve student wellbeing in the classroom–in-person or online–and protect their privacy while also ensuring they feel supported as a whole person. Since UDL is premised on student agency and choice, its implementation in courses can prevent the need for students with disabilities to activate their accommodations altogether. CRP goes one step further, helping students feel that their experiences and identities are not a hindrance but rather a unique and valuable perspective. That said, respecting students' choices and providing space to share rather than pressuring students to share is fundamentally important, as trauma-informed teaching demonstrates. Though not always the case, disabilities can stem from very traumatic events, so it is important to welcome students' perspectives without making them relive trauma or feel impelled to share that trauma in a classroom setting (Morrison & Casper, 2012). Cultivating a careful balance of accessible content, valuing students' life experiences, and allowing students choice regarding what they keep private and what they are comfortable sharing will help create a more robust and engaging learning environment for all.

A key cornerstone of UDL, of course, is to provide choice in the types of learning materials (engagement) and the methods of assessment (expression). For example, introducing a choice between reading a written text, watching a video, or listening to a podcast can allow students to select the option which works best for their own learning. A student with epilepsy may choose to avoid the video and instead listen to the podcast. They do not need to request an accommodation or reveal that they are unable to watch videos with certain visual stimulation, which may feel uncomfortable since it is a symptom easily linked to the condition. Similarly, providing choice in how students express or demonstrate their learning provides them with agency to choose the most appropriate way to show their comprehension. Instead of writing an essay, a student with learning disabilities which affect their written work might instead choose to present verbally. Hosting materials online for remote learning has made it easier for instructors to provide materials in multiple formats thereby increasing accessibility.

Another key form of choice that benefits students with disabilities is flexibility with deadlines. While a timeline is important, especially when assignments are designed to scaffold or build on previous work, students can benefit from clear policies which allow limited extensions. For example, clarifying in the syllabus that while a specific deadline is provided, students can take up to an additional week to submit the work without penalties or need to contact the instructor can mitigate burden on students to activate accommodations. In the case of McMaster University's SAS accommodations, consideration for up to a week of extension on an assignment deadline is a common accommodation (McMaster University – SAS, 2022). That said, it requires students to contact an instructor in advance of the deadline to arrange and confirm the extension. Providing a blanket policy for the entire class can benefit instructors, who may no longer need to liaise with numerous students to negotiate individual accommodations for each assessment. The limited length of the extension keeps students close to being on track and can also spread out the burden of grading, providing space between assessments' submission.

UDL also proposes that instructors should vary the means of representation, which aligns well with culturally-responsive teaching practices. Though individual instructors cannot change broader societal and systemic biases against individuals with disabilities, they can address their own classroom environments and create space for students' contributions from their lived experiences. Beyond eliminating the need for disclosures which violate student privacy, CRP advocates including and welcoming culturally varied perspectives in the classroom, including disability perspectives. As Dufour explains, though disabilities (in her case, specifically learning disabilities) are not inherently the result of culture, students with disabilities have "cultural knowledge" which "stems from students' lived experience" and "presents opportunities for enhancing learning." Bringing representations of disability into course content can help students with disabilities in the classroom feel more confident that their perspectives and lived experiences are valuable (Dufour, n.d.).

On a practical scale, incorporating CRP for disability can be very simple. If providing case studies, particularly when visual aids are used, consider incorporating an individual with a disability. The purpose is not to call attention to any perceived limitations that that individual may face, but rather to show individuals with disabilities living in the world, as a natural component of society. Instructors may also consider incorporating work written by or created by individuals with disabilities where appropriate, which share their own perspectives on living with disability. These approaches are similar to educators' responses to calls to diversify reading lists and incorporate perspectives beyond traditional power-holders in society (MacPherson Institute, 2021). In addition to helping students with disabilities feel confident in their identities, the exposure to broad cultural perspectives is also beneficial to students without disabilities, who may not have engaged with, or been aware of engaging with, individuals with disabilities, particularly when those disabilities are invisible. Demystifying disability helps to destigmatize it.

While including students with disabilities and ensuring their needs are met is important, it is also beneficial to student wellbeing to avoid any perceived pressure to disclose. Disclosures can be traumatic for students and may impede their learning by inducing anxiety or even triggering post-traumatic stress disorder. Trauma-informed teaching strategies recognize that students may experience all sorts of trauma, including, but not limited to, violence or medical trauma which may cause disability. Indeed, the CDC reported in 2019 that 61% of adults surveyed across 25 states reported experiencing at least one form of "adverse childhood experience (ACE)" before the age of eighteen (Centers for Disease Control and Prevention, 2021). Though it would be incorrect to assume that all individuals with disabilities have experienced trauma related to those disabilities, educators need to acknowledge that for some, this may be the case. By including examples of individuals with disabilities in course content and cultivating an environment in which all students' perspectives and cultural currency are clearly valued, educators can cultivate space for students to share if they are comfortable. Shaping course content to minimize the need for students' disclosures in the form of accommodation requests is crucial, and it is important that educators do not replicate that pressure in the classroom. Again, though maintaining awareness of students' potential traumas is beneficial for students with disabilities, the problem of ACE is not restricted to this population. Thus, incorporating this teaching technique is broadly beneficial to students as well.

Though the term "confidentiality" is a hallmark of accommodations policies, students' privacy can be better protected by instructors, SAS, and the university more broadly. Inherently, the accommodation process at most institutions requires disclosure of a disability, whether visible or invisible. This act of disclosure may not include describing the specific type of disability, but it still requires students to prove that they need specific accommodations to achieve equitable learning conditions. In this model, the onus is placed on the individual student to ensure that they can access their education. An instructor may also need to provide accommodations of different types to many students in the class, requiring additional work on the instructor's part.

Many instructors, however, report feeling overwhelmed at the thought of having to completely redesign a course to ensure accessibility, equity, and inclusion. Nonetheless, the goal should be progress, not perfection. Improving accessibility, adding choice, and incorporating trauma-informed pedagogy and CRP is an ongoing process. Awareness of the possibilities and the benefits of applying these teaching practices to courses is crucial. Students' privacy is of fundamental importance and, as a by-product of protecting students with disabilities' privacy, those students and their peers can benefit from more diverse ways of knowing and more ways to learn and show comprehension. In addition, instructors may benefit from fewer accommodation requests, as fewer students will have difficulty accessing materials.

Conclusion

The incorporation of UDL, trauma-informed teaching, and CRP can support students with disabilities and limit their need to disclose personal circumstances to others. If educators consider how to improve flexible course design, inclusion and representation of disability in content, and respect for students' lived experiences, not only will students with disabilities require fewer accommodations, but all students will benefit. Limiting the time that students with disabilities spend liaising with SAS or its counterparts—perhaps necessitating extra visits to medical professionals and revisiting traumas in their past—and working to ensure educators implement the proper accommodations will leave students with more capacity to focus on their studies. Beyond simply advocating for equitable access to course content, CRP is a useful tool to ensure students with disabilities feel valued in the classroom, with the caveat that students should not feel impelled to share their conditions. Instead, students should be provided with multiple means of representation, ensuring that they can see themselves in course content, and that they have space to share their own perspectives, informed by their lived experiences. In this way, we can mitigate breaches of privacy and allow students to focus on learning.

References

Barr, D. A. (2018). When trauma hinders learning. *Phi Delta Kappan 99*(6), 39-44. https://doi.org/10.1177%2F0031721718762421

Basham, J. D., Israel, M., Graden, J. & Poth, R. (2010). A comprehensive approach to RTI: Embedding universal design for learning and technology. *Learning Disability Quarterly, 33*(4), 243-255. https://www.jstor.org/stable/23053228

Blanks, A. B., & Smith, J. D. (2009). Multiculturalism, religion, and disability: Implications for special education practitioners. *Education and Training in Developmental Disabilities, 44*(3), 295–303. https://www.jstor.org/stable/24233476

CAST. (2018). *Universal Design for Learning guidelines*. CAST. http://udlguidelines.cast.org

CAST. (2022). *About Universal Design for Learning*. CAST. https://www.cast.org/impact/universal-design-for-learning-udl

Cavanaugh, B. (2016). Trauma-informed classrooms and schools. *Beyond Behaviour 25*(2), 41-46. https://www.jstor.org/stable/26381827

Centers for Disease Control and Prevention. (2021). *Adverse childhood experiences prevention strategy*. National Center for Injury Prevention and Control, Centers for Disease Control and Prevention. https://www.cdc.gov/injury/pdfs/priority/ACEs-Strategic-Plan_Final_508.pdf

Cook, S. C., & Rao, K. (2018). Systematically applying UDL to effective practices for students with learning disabilities. *Learning Disability Quarterly 41*(3), 179-191. https://doi.org/10.1177%2F0731948717749936

Coughlan, T., Lister, K., Seale, J., Scanlon, E., & Weller, M. (2019). Accessible inclusive learning: Foundations. In R. Ferguson, A. Jones, & E. Scanlon (Eds.), *Educational visions: The lessons from 40 years of innovation* (pp. 51-73). Ubiquity Press. https://doi.org/10.5334/bcg.d

Dufour, Eve. (n.d.). Learning disabilities and diversity: A culturally responsive approach. *LD@School*. https://www.ldatschool.ca/culturally-responsive-pedagogy/

eCampusOntario. (2022). eCampusOntario Programs. https://search.ecampusontario.ca/itemTypes=2&sourceWebsiteTypes=1&returnUrl=https://ecampusontario.ca/learners/#/?sortCol=2

Gallagher, D. J., Connor, D. J., & Ferri, B. A. (2011). Broadening our horizons: Toward a plurality of methodologies in learning disability research. *Learning Disability Quarterly, 34*(2), 107–121. https://www.jstor.org/stable/23053255

Gay, G. (2000). *Culturally responsive teaching: Theory, research, and practice.* Teachers College Press.

Gay, G. (2002). Preparing for culturally responsive teaching. *Journal of Teacher Education, 53*(2), 106-116. http://dx.doi.org/10.1177/0022487102053002003

Human Resources and Skill Development Canada. (2010). *Federal disability report: The Government of Canada's annual report on disability issues.* Government of Canada. https://publications.gc.ca/collections/collection_2010/rhdcc-hrsdc/HS61-1-2010-eng.pdf

Johnson-Harris, K. M., & Mundschenk, N. A. (2014). Working effectively with students with BD in a general education classroom: The case for universal design for learning. *The Clearing House: A Journal of Educational Strategies, Issues and Ideas 87*(4), 168-174. https://doi.org/10.1080/00098655.2014.897927

McKenzie, C.. (2015). Navigating post-secondary institutions in Ontario with a learning disability: The pursuit of accommodations. *Canadian Journal of Disability Studies 4*(1), 35-58. https://doi.org/10.15353/cjds.v4i1.186

McMaster University – Student Accessibility Services (SAS). (n.d.). *Guidelines for the provision of documentation for students with disabilities.* https://sas.mcmaster.ca/wp-content/uploads/2022/06/SAS-Medical-Documentation-April-2017.pdf

McMaster University. (2020). *Academic accommodation of students with disabilities.* https://secretariat.mcmaster.ca/app/uploads/Academic-Accommodations-Policy.pdf

Michael, M. G., & Trezek, B. J. (2006). Universal design and multiple literacies: Creating access and ownership for students with disabilities. *Theory Into Practice 45*(4), 311-318. https://www.jstor.org/stable/40071615

Morrison, D., & Casper M. J. (2012). Intersection of disability studies and critical trauma studies: A provocation. *Disability Studies Quarterly 32*(2). https://dsq-sds.org/article/view/3189

Olney, M., & Brockelman, K. (2003). Out of the disability closet: Strategic use of perception management by select university students with disabilities. *Disability & Society 18*(1), 35-50. https://doi.org/10.1080/713662200

Ontario Human Rights Commission. (2002). *The opportunity to succeed. Achieving barrier-free education for students with disabilities.* Ontario Human Rights Commission https://www.ohrc.on.ca/en/opportunity-succeed-achieving-barrier-free-education-students-disabilities/post-secondary-education

Paul R. MacPherson Institute for Leadership, Innovation and Excellence in Teaching & Equity and Inclusion Office (MacPherson Institute). (2021).

McMaster's inclusive teaching and learning guide. McMaster University, Hamilton, Ontario. https://mi.mcmaster.ca/inclusive-teaching/#tab-content-ov

Rose, D. H., Harbour, W. S., Johnston, C. S., Daley, S. G., & Abarbanell, L. (2006). Universal design for learning in postsecondary education: Reflections on principles and their application. *Journal of Postsecondary Education and Disability, 19*(2), 135-151.

Scott, A. N., Boynton Hauerwas, L., & Brown, R. D. (2014). State policy and guidance for identifying learning disabilities in culturally and linguistically diverse students. *Learning Disability Quarterly 37*(3), 172–85. https://doi.org/10.1177/0731948713507261

Sitler, H. C. (2009). Teaching with awareness: The hidden effects of trauma on learning. *The Clearing House, 82*(3), 119-123. *JSTOR*, http://www.jstor.org/stable/30181092.

Taylor, K. (2016). The influence of universal design for learning through cooperative learning and project-based learning on an inclusive classroom. *Student Research Submissions.* 187. https://scholar.umw.edu/student_research/187/

University of Guelph. (2016). *Academic accommodations for students with disabilities.* https://www.uoguelph.ca/secretariat/policy/2.1#:~:text=%22Academic%20Accommodation%22%20means%20modification%20to,the%20student's%20Disability%2Drelated%20needs.

University of Saskatchewan – Access and Equity Services. (2021). *Duty to accommodate procedures.* https://policies.usask.ca/documents/duty-to-accommodate-procedures.pdf

Vavrus, M. (2008). Culturally responsive teaching. In T. L. Good (Ed.), *21st century education: A reference handbook* (vol. 2, pp. 49-57). Sage Publishing. http://doi.org/10.4135/9781412964012.n56

Western University. (2019). *Western University policy on academic accommodation for students with disabilities.* Academic handbooks, rights and responsibilities, accommodation for students with disabilities. https://www.uwo.ca/univsec/pdf/academic_policies/appeals/Academic%20Accommodation_disabilities.pdf

Wilks, A. (2022). *Push Back on Back to Mac Panel Discussion* [Video]. MacVideo. https://www.macvideo.ca/media/Push+Back+on+Back+to+Mac+Panel+Discussion/1_bnbyb346

Remote Learning Environments for Students who are Academically at-risk, Non-traditional, or from Diverse Backgrounds

Christina M. Cobb and Meredith Anne (MA) Higgs

The Covid-19 pandemic brought severe challenges and unprecedented changes around the globe. In particular, the educational world had to shift quickly to accommodate the new normal for pandemic-era classes. The need for online and remote courses increased, and teachers everywhere were faced with trying to reach students in novel or non-traditional ways. Traditional face-to-face courses were no longer the best or only option. Teachers began to explore different remote platforms such as Zoom and Microsoft Teams, and these virtual formats were implemented across all levels of higher education. Therefore, the Covid-19 pandemic and the ensuing changes to higher education were challenging for students across academic disciplines and degree types; however, the virus and its aftermath were perhaps not fair in their disparate treatment of some higher education student populations. Indeed, some persons may have experienced the pandemic with greater concerns for personal safety and privacy than others.

While the pandemic was exceedingly difficult for many students, some may have experienced additional barriers to higher education attendance. For instance, numerous students worked in industries deemed essential, and essential employees were required to work during the initial waves of the pandemic. Students from diverse backgrounds or in certain occupations may have experienced higher likelihood of virus exposure or contraction (Hawkins, 2020) in industries such as health care (Nguyen et al., 2020) and food processing as was well-documented in many high-profile outbreaks (Donahue et al., 2020; Rogers et al., 2020). Leonhardt (2022) states that the pandemic initially had a "disproportionate toll on Black and Latino Americans" (para. 1). Consequently, these students were then more likely to be required to quarantine for Covid-related exposures or to isolate for illness. In addition, essential employees experienced work schedule changes due to pandemic-related staffing issues, which also necessitated classroom experiences that were remote or online. Indeed, remote learning became the standard to address higher education. Therefore, while remote higher-education attendance was initially a standard practice for all instruction at many higher educational institutions early in the pandemic, it proved to become a necessity for some student populations.

Moreover, home spaces changed dramatically overnight. Many families began sheltering together to address childcare issues. Communal spaces became over-run with extra people, possessions, and remote work and school areas; multiple children and adults attempted to use the same technological resources at the same or similar times. This chaotic life was not only difficult to sustain, it could have been embarrassing when seen on Zoom by peers and teachers.

This chapter introduces some of the issues and successes that remote teaching brought during the pandemic, as well as continual issues for students who are academically at-risk, non-traditional, or from diverse backgrounds.

Definitions

For the purposes of this chapter, the following working definitions are used in describing the impact of remote learning environments on students who are academically at-risk, non-traditional, or from diverse learning backgrounds.

Students who are academically at-risk are defined as students with academically deficient backgrounds or with past academic histories that make them more likely to be unsuccessful. The term "at-risk" has challenges; however, it has been long used to describe students who are more likely to fail (National Center for Educational Statistics, 1992).

Students who are non-traditional are learners who are over the age of 24 years and/or have family and work responsibilities that can complicate higher educational attainment (National Center for Educational Statistics, n.d.).

Students who are from diverse backgrounds are learners who have "racial, cultural, and/or life experiences" that are different from the instructor (Nishioka, 2018, para. 4).

Disparate implications is an author-derived term describing the unintentional, unequal consequences of educational policies or activities.

Camera usage refers to the videographic or imaging technologies required for instructional participation including live video and image recording.

Test proctoring describes the various means of monitoring students while they are taking examinations. These include methods such as live video observations, body movement monitoring, lock-down browsers, and remote computer control.

Private spaces is an author-derived term describing the homelife or living situations of students that are typically not engaged in the traditional classroom setting.

Cameras: The Good, the Private, and the Unequitable

Student Engagement and Cameras

The sudden move to remote and online coursework during the pandemic left many teachers scrambling to identify ways to keep their students engaged. In the remote and online environment, camera usage became the primary means of face-to-face interaction and a tool for student engagement. Truly, whether to require student cameras to be turned on during class meetings is a continuing conundrum for faculty (Torchia, 2021). Requiring cameras to be turned on during class time positively changes the culture of the classroom for students who are academically at-risk, non-traditional, or from diverse backgrounds (Racheva, 2018). By requiring cameras to be on, students are visible to faculty and peers in most videoconferencing platforms. Names, email, or nicknames are often visible as well and make it easier for students and faculty to identify each other. Also, faculty can better monitor student reactions to work and look for those moments of confusion that are the hallmarks of poorly described problems. According to Will (2020), despite the challenges with requiring students to turn on their cameras, teachers find that it is easier to check to see if students are participating, following the instructional content, or looking puzzled. With cameras turned on, students get to know each other and can form a culture that encourages further engagement and forges community. In addition, knowing that others are watching, and perhaps recording, may change student behavior and encourage students to stay on-task. Instructors also may find it easier to remember students who are visibly present rather than a stationary caricature or photograph.

As stated by Raicu (2020), to help build community in an online environment, faculty members should educate their students on the multidimensional need for authentic community. In doing so, students can see the real benefits for building community even in a remote or online setting. In a study done by Bedenlier et al. (2021), peer-to-peer interaction was an identified issue because students may feel less social support which may cause them to be less engaged in the course. Therefore, when students feel as though they belong and have social interaction with their peers, they are more likely to feel comfortable engaging in the course. Taken together, engagement, interaction, community-building, and classroom culture are compelling reasons for faculty to require student camera usage in remote environments.

Challenges with Cameras: Private and Unequitable

Although there are important student engagement-related justifications for requiring cameras to be turned on, there are also some inhibiting considerations for students. Factors such as Zoom fatigue (Moses, 2020) may persist in remote and online environments, especially during prolonged periods like the Covid-19 pandemic. Zoom fatigue, or fatigue associated with any videoconferencing platform, is when students are exhausted or experience burnout from the overuse of video platforms. Moses (2020) states that although most think that Zoom fatigue is no different from routine educational fatigue, there is a difference, and continuous video meetings may intensify the issue. Having cameras turned on during remote and online courses can facilitate engagement, but teachers should also consider the possible negative consequences of continual video usage.

The reliance on new formats of technologies increased as the world of higher education shifted to virtual classes, and this shift is now seen as the new custom for many classes. While this was very helpful to ensure that students were still being taught during the crisis phase of the pandemic, it also brought on challenges for some students. Dutta et al. (2020) states that these "digital spaces reify and reproduce ongoing inequalities" in addition to the disparities that Covid-19 also reproduced (p. 18).

During the pandemic, not only was access to technology a challenge, but the issue of cameras in students' private spaces also arose. Truly, engagement comes at several costs. As previously mentioned, students who are academically at-risk, non-traditional, or from diverse backgrounds may have complicated home situations, may be using technology from a free internet source, such as a restaurant, or may have to show private spaces that are embarrassing. Traditionally, students were not required to reveal any information about their personal lives but requiring cameras to be turned on invades that shield of privacy (Moses, 2020). With a focus on creating equitable and inclusive classrooms, teachers were faced with the dilemma of asking and/or requiring camera usage during class. While this may seem a small consideration, to a student with a complicated home environment, turning on the camera could be embarrassing or seen as a source of anxiety.

In a study by Castelli and Sarvary (2021), surveyed students responded as to why they did not turn on their cameras during class. The study revealed that 41% of respondents were "concerned about their appearance" while 26% were worried about "other people being seen in the background" (p. 3568). In a study done by Tobi et al. (2021), a lack of quality internet connection was the highest-ranked reason for cameras to be turned off during class. Students' access to stable internet service is a major concern that teachers should remember when requiring cameras to be turned on during their remote and online courses.

Another worrisome consideration is students who lack social skills and behavioral norms and who may then over-display these inappropriate behaviors in remote and online environments. In a world where some students post every thought and action, it may be difficult for those students to discern what is and what is not appropriate to either share with or record from others. Students may be concerned that their peers are going to use their class responses as opportunities to record the newest TikTok or create an internet meme. Students with children may be further concerned for their children's safety if inadvertently recorded in class. Not surprisingly, when cameras are required to be turned on, students' feelings of distress and nervousness can be intensified during their remote and online courses (The Sheridan Center, n.d.).

Finally, Trust (2020) suggested that educators should be trained on how to evaluate technologies for the classroom because they may be unknowingly violating students' privacy rights. These violations may, in turn, put students in dangerous or exploitative situations, such as providing personally identifiable information over an internet-based technology, showing students' location details, and sharing various computer usage information. Creating dangerous or exploitative situations should, most assuredly, never be the intent or consequence when utilizing technologies to enhance the higher education classroom. Such potential violations of privacy and security can create additional anxiety and stresses for students and raise equity issues.

Test Proctoring

Test proctoring platforms often use behavioral algorithms, recorded sessions, computer control, or observers to monitor testing. Obviously, many of the same considerations given to camera usage apply to test proctoring. Moreover, as many test recordings are stored off-campus, students may have additional safety and security concerns for themselves and their families. Students may be more concerned utilizing third-party companies' products as opposed to products from higher education institutions (Levy et al., 2011). Normal student behaviors may also be an issue. Test anxiety in proctored online exams is not well studied and could impact student success (Woldeab & Brothen, 2019). In academic support courses, which are often required for students who are academically at-risk, non-traditional, or from diverse backgrounds, test anxiety may especially be a concern. Using the test proctoring platforms may add to technological concerns for students as the platforms could require specific computer processing speed, camera quality, microphone use, and strong bandwidth. Household sharing of computers may have resulted in a physical lack of resources while multiple devices using the same internet connection could compromise bandwidth (Richards et al., 2021). Not understanding the security features or behavioral expectations of the testing platform may also unfairly target students who have never been exposed to the technologies. All things considered, test proctoring platforms can add barriers to success for remote or online students who are academically at-risk, non-traditional, or from diverse backgrounds.

Best Practices for Safety and Security

Higher education faculty can take basic steps to aid in student safety and security in remote and online environments while still fostering student engagement. Encouraging students to use their cameras whenever possible can help build community, foster classroom culture, and increase engagement. Faculty should consider surveying the class to determine who may have any technological issues and if they prefer to turn on their cameras during class. This survey will allow faculty members to manage their next steps in creating a classroom that is inclusive and fair for all.

Also, to prevent student embarrassment, students should be encouraged to utilize appropriate backgrounds or the blur function as well as to be given the option to leave the camera off when needed. Specifically, in Zoom, faculty members may utilize the "ask to turn on camera" option. This gives students the option whether to turn on their cameras, while still demonstrating that camera use is preferable. Another alternative is to provide backgrounds that are institution-specific. Many students will require demonstrations on how to use and upload backgrounds.

One very important step is to have specific dialogue and syllabus statements concerning safety, security, camera usage, and class recording. Specifically, instructors should create and implement a "cameras on" policy for the course syllabus. This will inform students from the start of the course what the class expectations are for camera usage. Faculty members may consider providing in-class and out-of-class sessions that address safety and security settings of technological platforms and devices. Instructors should promote access to safe campus technological resources and technical support options. Truly, protecting the most at-risk students is vitally important. By providing both synchronous and asynchronous instructional options, students who are most impacted by remote and online instruction can access content when, where, and how they feel most safe. Further, for test proctoring, consider utilizing instructor-proctored opportunities to reduce the disparate implications of lack of technological resources.

Perhaps the most basic and most useful strategy is building a rapport with students in a remote or online course from the very first opportunity. As repeatedly demonstrated in higher educational literature, utilizing High Impact Practices (HIPs) such as collaborative assignments and projects (Kuh et al., 2017) can aid in increasing engagement. Increased student engagement may, in turn, aid in the development of a culture of safety and security. Consequently, students may be more forthcoming with all types of issues and concerns. Therefore, building a classroom culture with clearly delineated behavioral and community expectations will enhance engagement, comfort, safety, and participation.

Conclusion

Since the beginning of the pandemic, the world of higher education has transitioned into a more diverse learning environment with additional remote or online learning opportunities. Early in the pandemic, traditional face-to-face courses were transformed into technology-based courses. During this rapid transition, instructors were faced with developing new techniques for teaching remotely. Some of those techniques included using platforms such as Zoom and Microsoft Teams to attempt to create an environment that was as engaging as traditional face-to-face classes. However, with use of these new platforms came new safety and security concerns and equity issues. Indeed, safety, security, and equity issues abound in remote and online instructional formats for students who are academically at-risk, non-traditional, or from diverse backgrounds.

Remote and online instructors are faced with novel challenges and must work towards making all their classrooms inclusive, equitable, and safe. Having cameras turned on in the students' private spaces was and still remains a conundrum. Although the world of higher education is attempting to return to pre-pandemic normalcy, many classes have returned to a traditional in-person format. However, there are still many courses that employ technology-based platforms for instruction and are continuing to wrestle with the issue of cameras in students' private spaces. As mentioned in the best practices section, offering students alternatives during class, such as the use of a virtual background or the blur setting, can ease some of these concerns.

As we strive to improve student engagement in the online and remote settings, we should continue to remember that students who are academically at-risk, non-traditional, or from diverse backgrounds may already be dealing with outside stressors (Higgs et al., 2021). Fullan (2020) states that "Covid-19 and its associated pandemic exposed more explicitly great inequalities such as access to devices, platforms, and/or places to do schoolwork outside schools in education systems" (p. 26). Surveying students on or before the first day of class to consider their individual technological issues and needs may open avenues to enhanced dialogue and engagement. This survey information may inform the creation and implementation of camera-use policies that are fair for all students.

While institutions and faculty push for technology implementation, "the question of how to create a learning environment where all voices are equitably empowered in a broader context of technological disparity" ought to be at core of this push (Workneh & Lin, 2021, p. 499). Remember, the overall goal for any higher education faculty member is for all students to be successful. Ensuring that the class is meeting the needs and concerns of all students, no matter their characteristics, in remote and online environments can aid in student success. To support success and survival, institutions must adjust to the everchanging individualities of student populations (Higgs et al., 2020). Addressing the issues and challenges of students in remote learning environments is one area that needs continuing consideration, especially for students who are academically at-risk, non-traditional, or from diverse backgrounds.

References

Bedenlier, S., Wunder, I., Gläser-Zikuda, M., Kammerl, R., Kopp, B., Ziegler, A., & Händel, M. (2021) Generation invisible? Higher education students' (non)use of webcams in synchronous online learning. *International Journal of Educational Research Open, 2*, 1–8. https://doi.org/10.1016/j.ijedro.2021.100068

Castelli, F. R., & Sarvary, M. A. (2021). Why students do not turn on their video cameras during online classes and an equitable and inclusive plan to encourage them to do so. *Ecology and Evolution, 11*(8), 3565–3576. https://doi.org/10.1002/ece3.7123

Donahue, M., Sreenivasan, N., Stover, D., Rajasingham, A., Watson, J., Bealle, A., Ritchison, N., Safranek, T., Waltenburg, M. A., Buss, B., & Reefhuis, J. (2020). Notes from the field: Characteristics of meat processing facility workers with confirmed SARS-CoV-2 infection – Nebraska, April-May 2020. *MMWR: Morbidity & Mortality Weekly Report, 69*(31), 1020–1022. http://dx.doi.org/10.15585/mmwr.mm6931a3

Dutta, M. J., Moana-Johnson, G., & Elers, C. (2020). COVID-19 and the pedagogy of culture-centered community radical democracy: A response from Aotearoa New Zealand. *Journal of Communication Pedagogy, 3*, 11–19.

Fullan, M. (2020). Learning and the pandemic: What's next? *Prospects, 49*(1-2), 1–4. https://doi.org/10.1007/s11125-020-09502-0

Hawkins, D. (2020). Differential occupational risk for COVID-19 and other infection exposure according to race and ethnicity. *American Journal of Industrial Medicine, 63*(9), 817–820. https://doi.org/10.1002/ajim.23145

Higgs, M. A. S., Cobb, C. M., & Morris, P. D. (2020). Pre-COVID-19 on-ground and distance learners' characteristics: A road map for distance and remote learning during a pandemic. *Journal of Student Success and Retention, 6*(1), 1–15. http://www.jossr.org/wp-content/uploads/2020/11/Pre-COVID-19-On-ground-and-Distance-LearnersCharacteristics.pdf

Higgs, M. A., Cobb, C. M., & Morris, P. D. (2021). Active learning, students who are academically at-risk, and institutional classification. *Journal of College Academic Support Programs, 4*(1), 26–36. https://doi.org/10.36896/4.1fa2

Kuh, G., O'Donnell, K., & Schneider, C. G. (2017). HIPs at ten. *Change: The Magazine of Higher Learning, 49*(5), 8–16. https://doi.org/10.1080/00091383.2017.1366805

Levy, Y., Ramim, M., Furnell, S. M., & Clarke, N. L. (2011). Comparing intentions to use university-provided vs vendor-provided multibiometric authentication in online exams. *Campus-Wide Information Systems, 28*(2), 102–113. http://dx.doi.org/10.1108/10650741111117806

Leonhardt, D. (2022, June 9). COVID and race. *The New York Times*. Retrieved November 2, 2022, from https://www.nytimes.com/2022/06/09/briefing/covid-race-deaths-america.html

Martirosyan, N. M., Saxon, D. P., & Skidmore, S. T. (2021). Online developmental education instruction: Challenges and instructional practices according to the practitioners. *Journal of College Academic Support Programs, 4*(1), 12–23. https://doi.org/10.36896/4.1fa1

Moses, T. (2020, August 17). 5 reasons to let students keep their cameras off during zoom classes. *The Conversation*. https://theconversation.com/5-reasons-to-let-students-keep-their-cameras-off-during-zoom-classes-144111

National Center for Educational Statistics. (n.d.). *Nontraditional undergraduates/Definitions and data*. https://nces.ed.gov/pubs/web/97578e.asp#:~:text=A%20student%20who%20did%20not,school%20completion%20was%20considered%20nontraditional

National Center for Educational Statistics. (1992). *Characteristics of at-risk students in NELS:88*. https://nces.ed.gov/pubs92/92042.pdf

Nishioka, V. (2018, May 11). Building connections with students from diverse cultural backgrounds through perspective-taking. *REL Northwest*. https://ies.ed.gov/ncee/edlabs/regions/northwest/blog/building-connections.asp

Nguyen, L. H., Drew, D. A., Graham, M.S., Joshi, A. D., Guo, C. G., Ma, W., Mehta, R. S., Warner, E. T., Sikavi, D. R., Lo, C. H., Kwon, S., Song, M.,

Mucci, L. A., Stampfer, M. J., Willett, W. C., Eliassen, A. H., Hart, J. E., Chavarro, J. E., Rich-Edwards, J. W.,...Chan, A. T. (2020). Risk of COVID-19 among front-line health-care workers and the general community: A prospective cohort study. *The Lancet, 5*(9), e475–e483. https://doi.org/10.1016/S2468-2667(20)30164-X

Racheva, V. (2018). Social aspects of synchronous virtual learning environments. *AIP Conference Proceedings, 2048*(1), 020032. http://doi.org/10.1063/1.5082050

Raicu, I. (2020, August 25). *Immersive vs. intrusive learning: Cameras and fairness in online classes.* Markkula Center for Applied Ethics at Santa Clara University. https://www.scu.edu/ethics-spotlight/the-ethics-of-going-back-to-school-in-a-pandemic/immersive-vs-intrusive-learning-cameras-and-fairness-in-online-classes/

Richards, E., Aspegren, E., & Mansfield, E. (2021, February 4). *A year into the pandemic, thousands of students still can't get reliable wifi for school. The digital divide remains worse than ever.* USA TODAY. https://www.usatoday.com/story/news/education/2021/02/04/covid-online-school-broadband-internet-laptops/3930744001/

Rogers, T. M., Robinson, S. J., Reynolds, L. E., Ladva, C. N., Burgos-Garay, M., Whiteman, A., & Budge, H. (2022). Multifaceted public health response to a COVID-19 outbreak among meat-processing workers, Utah, March-June 2020. *Journal of Public Health Management and Practice, 28*(1), 60–69. https://doi.org/10.1097/phh.0000000000001383

The Sheridan Center. (n.d.). *Inclusive strategies for student camera use during zoom class sessions.* The Harriet W. Sheridan Center for Teaching and Learning at Brown University. https://www.brown.edu/sheridan/inclusive-strategies-student-camera-use-during-zoom-class-sessions

Tobi, B., Osman, W. H., Abu Bakar, A. L., & Othman, I. W. (2021). A case study on students' reasons for not switching on their cameras during online class sessions. *International Journal of Education, Psychology and Counseling, 6*(41), 216–224. https://doi.org/10.35631/IJEPC.641016

Torchia, R. (2021, July 20). *The pros and cons of requiring students to turn on their cameras.* EdTech Focus on K-12. https://edtechmagazine.com/k12/article/2021/07/pros-and-cons-requiring-students-turn-their-cameras

Trust, T. (2020, April 2). The 3 biggest remote teaching concerns we need to solve now. *EdSurge.* https://www.edsurge.com/news/2020-04-02-the-3-biggest-remote-teaching-concerns-we-need-to-solve-now

Will, M. (2020, October 20). *Most educators require kids to turn cameras on in virtual class, despite equity concerns.* EducationWeek.

https://www.edweek.org/teaching-learning/most-educators-require-kids-to-turn-cameras-on-in-virtual-class-despite-equity-concerns/2020/10

Woldeab, D., & Brothen, T. (2019). 21st century assessment: Online proctoring, test anxiety, and student performance. *International Journal of E-Learning & Distance Education, 34*(1), 1–26.

Workneh, T. W., & Lin, M.-C. (2021). Teaching global communication during covid-19: Challenges, mitigation, and lessons learned. *Journalism & Mass Communication Educator, 76*(4), 489–502. https://doi.org/10.1177/10776958211026176

Part 4: What Tools or Resources Can We Use?

"Have You Seen My Cartoon Yet?": Objectives on Managing Student Projects in an Online STEAM Program

Kristen Vogt Veggeberg

The COVID-19 pandemic hit multiple industries hard, including those in out-of-school learning environments. This was especially true for programs within science, technology, engineering, arts, and math (STEAM) for elementary and middle school-aged youth. As programs are normally done within a group or classroom setting, the COVID-19 pandemic made it impossible for these programs to meet in person. If STEAM programs were not shut down or canceled outright, they were quickly transferred to an online platform, but only if the focused subject was able to be transferred online. The shift to an online platform included but was not limited to Zoom-based courses led by instructors, downloadable video content, or social media live streams hosted by different platforms. The change of communication mediums allowed for a loose version of a "class" that mitigated the need for youth participants to meet in person but still allowed them to participate in STEAM programming. Eventually, many of these programs, ranging from dance, science labs, and chess (Liao, 2016), moved to become more in-person again in 2021, with distance modes in place to protect individuals.

For many within STEAM education, this program change included the application of transferring many of their offerings online as well. While some programs sought out follow-along guides, others focused on using online technology-focused platforms to teach STEAM materials, such as coding using code.org and Robotify to teach robotics (Kalelioğlu, 2015). Other mediums that could incorporate STEAM pillars like design and inquiry-based learning became more popular at this time, such as animation, video making, and cybersecurity (Hsiao et al., 2021). All of these followed a similar pattern and delivery of traditional programming, except for their readiness in case of a sudden transition to online learning.

Several programs continued this mode of delivering education in 2021 due to the uncertainty of rising pandemic numbers. Many institutions could not afford losing out on another year without providing programming. As a result, the changes implemented in the first year of the pandemic remained. Teachers needed to adjust online programs as necessary, where issues regarding privacy, safety, and security for the participants were involved. Incorporating cloud-based platforms was necessary not only to maintain education materials but also to preserve student projects. The following section will address how understanding the issues involving student privacy, safety, and security with these platforms is manifold.

The Use of Cloud-Based Platforms in STEAM Education

STEAM education can take many different forms, as its modalities cover a wide variety of subjects meant to engage with design and inquiry-based learning (Hawari & Noor, 2020). Subjects that are usually covered include but are not limited to robotics, chemistry, mobile app design, and drones. Instructional materials and guides for these subjects often focus on completing a final product at the conclusion of the lesson. Given the subject areas, many of these programs are inclined towards the use of technology, especially computers, and online work. For many STEAM educators, the transition to even more computer-based activities was an accepted part of many pedagogies during the COVID-19 pandemic, and the lessons learned during the pandemic regarding online education will perhaps stay permanent within computer science education. This shift to technology (the 'T' in STEAM) for use in multiple platforms of online environments is not a recent development. In fact, many of the subjects covered within STEAM often trace back to the computer itself, a combination of both technology education trends, and the pathway seen to technology within education (Shatunova et al., 2019). Computers play a role in why many out-of-school activities and programs choose technology-heavy components, including cybersecurity, where students work by using passwords. Because of the consistent use of computers within programs, there has been a somewhat easy transition to online learning in informal STEAM education, as a system was already set in place to engage with this form of technology within learning.

Educational activities regarding computing within STEAM often happen through multiple platforms and mediums, such as animation, coding (using different languages, such as Python), and designing both graphics and apps for a tablet. A storage platform is necessary for students and teachers to save these works and have them accessible across multiple modalities, such as a laptop, tablet, or cell phone. If teachers and students have an internet connection and a browser compatible with their digital device, cloud-based storage fits their requirements for preservation and storage.

Additionally, many of the practical modes of saving student work must be done on a cloud-based platform in the modern online classroom. In previous years, other formats have been used in lieu of cloud-based storage, such as a USB drive. However, this form of data management can come under scrutiny, especially when used by underaged youth in either social or academic settings. Physical forms of data management can be lost or destroyed easily, presenting a challenge for presentations, grading, or other forms of assessment. As a result, many instructors find a cloud-based storage platform for student work beneficial. Cloud-based storage is not an anathema to the challenges presented with having adequate data storage, especially within the application of protecting student identity and intellectual property online.

Challenges to Privacy and Security with Physical and Cloud-Based Storage

Cloud-based computing has become a consistent form of data management across multiple disciplines, especially in an increasingly online world (Ercan, 2010). Previously, storing data was done in individual hardware, such as floppy disks and CDs, both of which are hard to find in current educational technology. Some hardware, such as computers and USB drives, are still being used in a certain amount to this day, especially when the curriculum involves using a non-digital connection, such as a 3D printer (Hamidi et al., 2017). The popularity of these physical forms of data storage has been beginning to wane, due to both the ease of cloud storage and the impact of not being able to readily access a file if a physical copy is lost or damaged.

Cloud-based storage, one of the most popular formats in terms of storage of digital material and media, is often used through an accessible online system, such as Google Drive (Kirayakova, 2017). This free program is one example, though other examples are DropBox and a school's online platform. However, as this involves out-of-school learning, many organizations, especially those affected by the changing budgets due to the COVID-19 pandemic, must use free online platforms, namely Google Drive. This is not only because of the ease of access for many students but also due to the budget of the programs as well. Google products, which, while free to use for all educators and students, still collect valuable data from individuals, therefore putting privacy at risk.

File Size and Data Protection

Different data and file formats are used within STEAM education, and as a result, the physical storage has been kept around even when other environments that use technology have abandoned them. As of 2022, this ranges from files with small amounts of data (such as GIFs, commonly used in graphic design), to ones that can take up large swathes of data (namely stereolithography (STL) or .stls files, commonly used in 3D printing). This presents a challenge concurrent to student privacy, in that these large files can often use up too much bandwidth, and an external format must be used for students to save larger projects.

One of the most pertinent issues regarding student online safety is the ability to keep these large files protected in all formats. This includes online files, which are especially vulnerable to cybercriminals. As files within STEAM education can also be transferable to technologies such as 3D printers, non-cloud-based storage may be used unless more online space is purchased. Since these files are large, they may have constraints that could end with the loss of student work—such as the conversion from a bitmap file to a GIF file in images, which would change the structure of the image and thus the student's work itself. This option of purchasing more cloud-based storage has generally not been reserved for many STEAM programs during the COVID-19 pandemic, due to additional budgeting that may not be available for cash-strapped programs during the pandemic. But this difference in storage does not allow the amount of guaranteed backup in the instance of the students' files becoming lost or damaged if something happens to the physical storage component.

Cloud-based storage presents an additional threat to student safety. Keeping student data online in a cloud-based format leads to potential hacking through the use of unsafe or untrustworthy passwords by outside parties. The potential is that student data–including created files from the students themselves–can potentially be downloaded, shared, or otherwise exploited. This includes the intellectual property of students themselves, which has been proven to have issues regarding protection. Minors, unlike adults, often are not allotted the same intellectual property rights while completing their education. In addition to this issue, the challenge of making sure student identities are kept safe is also a present issue, even more so than protecting student creative materials.

As private information can be stored on cloud platforms, the abilities of teachers, administrative professionals, and even the students themselves, need special care in the application of online privacy. One format of this is the application of an authenticator, which is discussed in further detail below. Another is password safety and making sure that students have been trained in both being alert and knowledgeable about safety online, including that of sharing passwords with peers and other individuals who might know the student. This is a policy issue that individual districts and institutions should be aware of and solve in the wake of potential breaches of safety with student information.

Ways to Move Forward

STEAM subjects require online storage, especially during the COVID-19 pandemic. As students are online, so are the artifacts and materials they design. The use of this cloud storage, however, leaves students at the mercy of potential online criminals (Wu et al., 2020). There are ways in which cloud platforms can still be used safely at a larger scale. The initial response is to move all of the program storage to a physical format, as stated above. While this seems to work, especially with the application of STL files and other files needed to be plugged via USB formats, it is not a solution that can be easily replicated to protect student privacy. This is due to the challenge seen within distance learning that many have been experiencing since 2020 during the COVID-19 pandemic. In lieu of physical storage, files stored in online cloud platforms can be used and safeguarded in two different formats: stronger passwords and double authenticator applications. Both of these forms of cybersecurity have been readily available and used by professionals within different business settings and can easily be applied within a school setting.

Stronger passwords are a general recommendation for protecting student materials online and have become part of training for students regarding online safety, such as Google's "Be Internet Awesome" (Seale & Schoenberger, 2018). This includes a password generated by a digital artificial intelligence (AI) connected to an educator's user account for a cloud storage unit. This can be seen as one of the safest formats in which passwords can be given to program participants. The reason for this is that youth passwords can be easily replicated or guessed by non-student parties (Kurpjuhn, 2015). As a result, multiple authentications must be completed to protect the online presence of youth involved in any online platform.

Password safety has arguably been one of the primary challenges in not only student privacy but also for anyone engaging in online platforms for all users (Bartoli et al., 2015). Although there have been multiple attempts by such organizations as Google and Microsoft to promote and teach about cybersecurity (Corradini & Nardelli, 2020), the ability to reproduce protected passwords is a common hacking technique. Even with the prerequisites of different password authenticators, including longer (12 characters or more), more elaborate passwords, there are still large amounts of break-ins to online accounts, including those of students. There have been studies about passwords and youth, where youth may share a password with a close friend to show a sense of trust and camaraderie (Van Ouytsel, 2021).

Passwords are a crucial but challenging aspect of online platforms for underage youth participants to engage in. This allows for the storage and protection of materials and lessons made by and for students. But because of this ease of logging in and loading files to the cloud platform, it is easy for multiple parties to access student materials. These can include many of the previously mentioned file types, which can become quite large and need to be changed. Because of this, there leaves vulnerability of access from different parties, some of which may be seeking student materials for nefarious purposes.

Therefore, to protect students and the materials they create, other tools in addition to passwords should be used in the case of online platforms, even for storage. Authentications have also become more popular to use for safety in online formats. This requires a quickly sent code of a few numbers to be pressed into an authenticator platform. The authenticator then sends a personalized code to the individual, allowing them a brief window where they can access the platform using an individualized number generated through an automatic algorithm. This format is arguably more secure than alternative password resets, which can be frustrating for busy users who do not have the skills to remember complicated passwords (Woods & Siponen, 2018). The use of an authenticator, however, should always stay within the hands of the immediate program educator to watch over student materials online. This allows protective measures to be used for student materials stored on a cloud to be safe for storage. As youth are still at risk for outside influence for nefarious purposes, as seen within multiple examples of program participants being tricked by online cybercriminals, additional protection should be kept in this case (Rithika & Selvaraj, 2013). Authenticators, being free alongside other online platforms such as Google Drive, are usually easy to use, require no training and can be connected to different accounts for all participants. Signs of a good authenticator include a quick turnaround time to use the code (and discard it), an easily accessible medium for the user to gain the code (such as a text message or mobile app), and being protected by a password known only to the individual using the authenticator.

Although many formats of file sharing and interest are a part of STEAM education, student safety—whether for intellectual rights and purposes or for their own selves—comes first before enrichment. With the COVID-19 pandemic, more programs have become online, and more students have run the risk of either being the target of cybercriminals or having their work taken by third parties (Sastre-Merino et al., 2020). This risk, though not heavily explored for nonadult participants within design, is something that should be taken into consideration with the application of online STEAM programs, cementing the need for keeping such items as passwords and authenticators in hand. Thus, these formats, both stronger passwords and authentication, are current solutions for challenges present within protecting online cloud platforms that have been used by students and their educators.

Conclusion

The range of locations in which students learn online has expanded during the COVID-19 pandemic, thus an online platform is present within these educational activities. In regard to access across multiple computers in different locations, there are still present issues in protecting student privacy. Issues regarding student safety with physical data have always been a concern, and one in which cloud computing has served throughout different places that require access. For many individuals and the organizations that they work with in STEAM education, digital learning has become a way of life. This need to use digital learning to protect students and continue education is true even as jurisdictions look to ease the population back into a pre-pandemic lifestyle. Protecting students and their digital artifacts is necessary for online learning environments. While cloud-based storage is a valuable tool in maintaining student progress and work, instructors must take precautions to protect student files, ensuring their safety and security while navigating these online spaces.

References

Bartoli, G., Fantacci, R., Gei, F., Marabissi, D., & Micciullo, L. (2015). A novel emergency management platform for smart public safety. *International Journal of Communication Systems, 28*(5), 928-943. https://doi.org/10.1002/dac.2716

Corradini, I., & Nardelli, E. (2020). Developing digital awareness at school: A fundamental step for cybersecurity education. In I. Corradini, E. Nardelli, and T. Ahram (Eds). *International Conference on Applied Human Factors and Ergonomics* (pp. 102-110). Springer. https://doi.org/10.1007/978-3-030-52581-1_14

Ercan, T. (2010). Effective use of cloud computing in educational institutions. *Procedia-Social and Behavioral Sciences, 2*(2), 938-942. https://doi.org/10.1016/j.sbspro.2010.03.130

Hamidi, F., Young, T. S., Sideris, J., Ardeshiri, R., Leung, J., Rezai, P., & Whitmer, B. (2017, May). Using robotics and 3D printing to introduce youth to computer science and electromechanical engineering. In *Proceedings of the 2017 CHI Conference Extended Abstracts on Human Factors in Computing Systems* (pp. 942-950). Association for Computing Machinery. https://doi.org/10.1145/3027063.3053346

Hawari, A. D. M., & Noor, A. I. M. (2020). Project based learning pedagogical design in STEAM art education. *Asian Journal of University Education, 16*(3), 102-111. https://doi.org/10.24191/ajue.v16i3.11072

Hsiao, P. W., & Su, C. H. (2021). A study on the impact of STEAM education for sustainable development courses and its effects on student motivation and learning. *Sustainability, 13*(7). https://doi.org/10.3390/su13073772

Liao, C. (2016). From interdisciplinary to transdisciplinary: An arts-integrated approach to STEAM education. *Art Education, 69*(6), 44-49. https://doi.org/10.1080/00043125.2016.1224873

Kalelioğlu, F. (2015). A new way of teaching programming skills to K-12 students: Code. org. *Computers in Human Behavior, 52*, 200-210. https://doi.org/10.1016/j.chb.2015.05.047

Kiryakova, G. (2017). Application of cloud services in education. *Trakia Journal of Sciences, 15*(4), 277-284. https://doi.org/10.15547/tjs.2017.04.001

Kurpjuhn, T. (2015). The SME security challenge. *Computer Fraud & Security, 2015*(3), 5-7. https://doi.org/10.1016/S1361-3723(15)30017-8

Rithika, M., & Selvaraj, S. (2013). Impact of social media on students' academic performance. *International Journal of Logistics & Supply Chain Management Perspectives, 2*(4), 636-640.

Sastre-Merino, S., Nuñez, J. L. M., Pablo-Lerchundi, I., & Nufiez-del-Rio, C. (2020, December). Training STEAM educators in the COVID-19 emergency situation: Redesigning teaching. In *2020 Sixth International Conference on e-Learning (econf)* (pp. 72-75). IEEE. https://doi.org/10.1109/econf51404.2020.9385461

Seale, J., & Schoenberger, N. (2018). Be Internet awesome: A critical analysis of Google's child-focused Internet safety program. *Emerging Library & Information Perspectives, 1*(1), 34–58. https://doi.org/10.5206/elip.v1i1.366

Shatunova, O., Anisimova, T., Sabirova, F., & Kalimullina, O. (2019). STEAM as an innovative educational technology. *Journal of Social Studies Education Research, 10*(2), 131-144.

Van Ouytsel, J. (2021). The prevalence and motivations for password sharing practices and intrusive behaviors among early adolescents' best friendships – A mixed-methods study. *Telematics and Informatics, 63*. https://doi.org/10.1016/j.tele.2021.101668

Woods, N., & Siponen, M. (2018). Too many passwords? How understanding our memory can increase password memorability. *International Journal of Human – Computer Studies, 111*, 36–48. https://doi.org/10.1016/j.ijhcs.2017.11.002

The Cost of Respect? Surprisingly Little

Joseph Kennedy and Albert Kagan

Online and distance learning classes have grown precipitously in the last decade. Flexibility, access, cost, class variety, and instructor engagement as well as technological advances have driven this expansion of online offerings, even while traditional class delivery across higher education is decreasing (Bailey et al., 2018; Snyder et al., 2019). During the fall 2020 semester, students enrolled in distance education courses in United States post-secondary institutions were 72.8% of all attendees; a year earlier this statistic was 37.2% (U.S. Department of Education [DoE], 2022 and U.S. DoE, 2021 respectively). Much of the increase in online offerings is related to COVID-19; however, recent U.S.-wide survey data indicates that students' desire for flexibility in course offerings is likely to cause many institutions to offer more online courses than they had pre-pandemic (Venable, 2022, p. 36).

As online and distance learning have increased, concerns regarding academic integrity have similarly grown. Some concerns revolved around the lack of in-person contact and the apparent ease of cheating. Numerous studies have demonstrated an appearance of cheating in online classes in disciplines as diverse as business, engineering, and nursing (Daffin & Jones, 2018; Dyer et al., 2020; Harton et al., 2019; Seife & Maxwell, 2020). Such results continue to demonstrate the pattern in students' attitudes regarding the nature of cheating in an Internet-connected world that King et al. identified in 2009 when they found that almost three-quarters of business students at their institution believed cheating is easier in online courses. Together, these findings indicated that students do not feel "cheating" is "cheating" when the instructor does not specifically define what constitutes cheating. This attitude may be due to the prevalence of websites that provide quick answers to questions often asked on tests, the ubiquity of resources that eliminate the need for students to learn and memorize basic facts, or a shift in learner conceptualization regarding what is truly worth remembering. Regardless of the cause, prior to the COVID-19 pandemic, there was a growing disconnect between instructors' expectations of student behavior on assessments, and students' understanding of which cheating behaviors were unethical (Braff, 2011; Seife & Maxwell, 2020).

When the COVID-19 pandemic started to spread in the winter and spring terms of 2020, many higher education intuitions adopted online tools rapidly. Oftentimes the hasty implementation was fraught with faculty/staff unpreparedness, technology deficiencies, limited course material availability, and student confusion. Coupled with the immediacy of the online transition were ongoing concerns regarding appropriate implementation of Universal Design for Learning (UDL) principles in online courses (Evmenova, 2018), integrity within the online model (Palmer et al., 2019), and a lack of clarity regarding the definition of academic integrity. Suryani and Sugen (2019) reported that it was difficult to locate the academic integrity policies of many institutions even pre-pandemic. Some schools that offered only a few courses online had a poorly developed technology infrastructure, and many of those schools were concerned about the cost of purchasing new technology tools in a short time span.

Meanwhile, students challenged new teaching methodologies; there was frustration regarding poor communication from faculty and institutions, confusion over differing proctoring methods, and considerable concern about student privacy. While some of the challenges could be addressed through clearer communication (Bozkurt et al., 2020), student concerns regarding privacy are pervasive, enduring, and multi-faceted. The anger of students revealed in one editorial (The Editorial Board, 2021) goes beyond resentment regarding the invasion of student privacy to argue that test proctoring software uniquely creates inequity. Other student voices articulate that such technologies are not only insulting and anxiety-producing, but fail to enhance academic integrity (Poster, 2021).

Traditional residential liberal arts institutions that did not already support a robust online class presence were particularly challenged as the pandemic became pervasive in the spring of 2020. The required training of faculty and students, course material development, grading alterations, workload modifications, technology upgrades, and enforcement of academic integrity standards challenged not just institutional capacity and budgets but also instructional models. The integrity standards had to address class applications, written assignments, group activities, and exam/assessment methodologies in an online setting. In this context, the paradigm of respect for learning, the students, the class, the instructor, and the institution had to be balanced with the concern for assessment security and the potential intrusiveness of any toolkit adopted by an institution.

This paper discusses the implementation of online assessment mechanisms using pre-existing tools at a small liberal arts college in the upper Midwest region of the United States, precipitated by the shift to distance learning driven by the COVID-19 pandemic. A set of online processes has been in place since the spring term of 2019, with continual refinement of online delivery methods, learning management system (LMS) practices, faculty training, and student instruction within a model of academic integrity preservation. The attainment of mutual respect across the system regarding faculty goals and student needs is a guiding principle supporting the institutional mission during the transition to online class delivery.

The Pandemic's Impact on Online Tool Usage: A Case Study

The authors' institution is a residential liberal arts college located in the upper Midwest with an approximate enrollment of 2,000 undergraduate students. The institution emphasizes in-person instruction and had only begun incorporating some hybrid and online courses into the curriculum in the prior four years. During the spring semester of 2020, 19 of 598 courses had been planned to be online or primarily online courses. The institutional use of educational technology tools to deliver performance assessments was relatively underdeveloped as well; less than one-third of courses taught in the fall of 2019 used quiz and test tools through the school's LMS. At this time, approximately 75% of the faculty members had no experience teaching courses including online experiences, and approximately one-half of the faculty members had limited experience using online assessment tools. These figures are based on course offerings at the institution from 2017-2020 supplied by the registrar's office, content available on the college LMS, and the semester-by-semester notes of one of the authors, who is the manager of LMS at the institution. This lack of familiarity with online course development is not isolated to this institution; multiple authors note limited online experience at many colleges in the spring of 2020 (Al-Freih, 2021; Cutri & Mena, 2020; Haslam et al., 2020; Johnson et al., 2020).

In late January 2020, this institution's classroom technologies coordinator, a member of Information Technology Services (ITS), argued that ITS needed to begin preparing the institution's infrastructure for the possible impact of COVID-19. His prescient warning allowed the department to procure some hardware ahead of the international rush and to begin preparing ITS staff to deploy and explain new technologies and learning approaches to the campus community.

During February of 2020, instruction continued to be in-person. Some faculty members began to inquire individually about tools and approaches for distance learning and sought advice from those faculty members who had prior experience with online courses. Meanwhile, the rapid advancement of COVID-19 led to a turbulent and ever-shifting set of policies; as of March 11, 2020, the institution had announced they had no plans to switch to online / distance learning. Two days later, the institution's president announced a new plan, which included a six-day "pause" in instruction, followed by a shift to fully remote learning. Given the institution's emphasis on face-to-face instruction, the institution's instructional designer felt this was an inadequate amount of time to prepare for online course continuation. This opinion was not unique to this college; other studies allude to the abruptness of the online conversion faced by many institutions (Dyer et al., 2020; Seife & Maxwell, 2020).

Development of Procedures to Safeguard Academic Integrity

Although faculty members had some concerns regarding academic integrity, much of the faculty focused on learning new techniques of teaching and the tools necessary to deliver instruction in a fully online environment. Therefore, relatively few faculty members engaged in discussions regarding academic integrity approaches and safeguards. Those who did engage in such discussions with the instructional designer identified the following issues, per his contemporaneous notes:

- Given the chaotic transition to online instruction, students would both likely feel increased pressure to violate academic integrity and would have greater opportunities to do so.
- The institution needed a system that provided confidence when a potential academic violation was identified so a Type I error would not occur; in other words, no students should be punished for having violated academic integrity when no such violation had occurred.
- It was important to avoid systems that were overly intrusive, which could lead to students feeling it was assumed they would cheat.
- The faculty wished to implement systems that were transparent and effective at deterring academic integrity violations.
- Systems that required students to use expensive equipment or tools would be inequitable, especially as the institution lacked sufficient funds to purchase hardware for each student.

Therefore, faculty members and administrators sought to balance effective deterrence and detection systems with a respect for student motivation and privacy. Given the residential face-to-face focus of the institution, faculty members needed to be acutely aware of the dangers of appearing overly intrusive. Thus, an early decision was made to avoid eye-tracking and non-college human proctoring systems. The desired balance also had to preserve the academic rigor of the courses, which meant that some systems of control and accountability were considered.

Unfortunately, with only one week to implement a radically different instructional model, faculty had little time to consider a holistic and institutional culture-based approach to academic integrity. While the institution did generally follow principles such as those articulated by Kitahara et al. (2011), namely that the problem of academic integrity violations must be addressed and solved at the societal level, the timeframe was clearly insufficient to determine the impact of new tool adoption on the shared cultural understanding of academic integrity. Instructors' focus on rapidly learning new tools and approaches meant they were often unable to clearly communicate procedures and objectives to students, as the faculty members themselves were unclear of the mechanisms being employed. As a result, University of Florida Instructional Assistant Professor D. Mani posits, the combination of the stress of the "unknown" and the breakdown in communication was likely one cause of resultant academic misconduct (personal communication, April 5, 2022).

Ultimately, several approaches were adopted. The authors implemented a secure-browser online assessment approach in the courses they taught and managed; during subsequent semesters, the authors managed ten courses with an approximate enrollment of 200 students. Eventually, this process would become a record-proctor model, including assessments delivered through the LMS using a secure online browser as well as video recording software originally designed as a performance assessment tool.

Additionally, the authors segmented the capstone paper for each class into three parts; expanded the number of questions in each quiz by five while allowing students to choose to skip any five questions; altered the participation mechanisms; and replaced letter/number indicators in multiple choice answer sets with mere bubbles. These changes were designed to provide greater support to students in line with UDL principles and demonstrated respect for their individual learning autonomy while also mitigating the risk of academic integrity violations. Breaking the capstone project into three segments allowed the instructor to provide rapid feedback, including redirection to students struggling with citation concepts as well as with content. It also minimized the value of any student's purchase of a "paper mill" submission. These refinements and the development of the recorded-proctoring assessment mechanism are described further in the following section.

Development and Maturation of this Design

The authors have been collaborating on course design and electronic tool incorporation in multiple courses offered in the School of Business since the fall of 2015; Course 1 was taught each fall and spring semester during the years discussed in this chapter. The original assessment design of the courses reflected Universal Design for Learning (UDL) considerations in multiple ways:

- Only two of the three mid semester exams were included in the semester grade calculation to ensure a student struggling on a particular day could still earn full marks.
- Most courses included a group presentation to allow students to demonstrate their knowledge and ability to operate in a group setting.
- Students added to their classmates' knowledge through both in-class participation and individual presentations covering current issues in the subject, allowing different modes of participation.
- The capstone (term) paper assignment was heavily scaffolded.

Fortuitously, the instructor had planned to offer courses online beginning in the spring of 2020, which were consequently modified mid-semester. In subsequent semesters, the courses were modified further as described below to reflect student feedback and to ensure that methods of instruction and assessment better demonstrated respect for students as learners as well as supporting non-intrusive monitoring practices.

Course 1

In the fall of 2019, Course 1 was offered in two sections as an in-person class. While the instructor used the institution's LMS to keep materials organized for students, the primary mechanism for demonstrating respect for students was face-to-face interaction, where the instructor could respond to stated and unstated student concerns in the moment. Relevant elements of this course included:

- 3 multiple-choice / essay mid-semester assessments, which were taken in person; students' lowest grade of these three assessments was dropped
- Study guides for each assessment
- A term paper with three defined parts, submitted as a single assignment near the end of the semester
- A group presentation of a case study, presented in person to the entire class
- An individual short presentation on a topic currently relevant to the course, during a week chosen by each student
- A participation grade, composed of interaction with student presentations and in-person attendance
- A final essay-based exam

In the spring of 2020, the instructor reimagined Course 1 as an online course, collaborating with the instructional designer to modify elements in a manner that would provide students greater autonomy without sacrificing the level of rigor nor impacting academic integrity. For each chapter, students were provided a study guide as well as a voluntary online quiz, using a format identical to the unit assessment. The unit assessments were now online tests with both multiple-choice and essay components; each student's lowest grade of the three was still discarded. In accordance with UDL principles regarding flexible assignments, students had a four-hour window to complete the assessment but were required to complete it within two hours. This flexibility again demonstrated respect for students but was not so extended in duration as to cause exam security concerns, consistent with the findings of Cluskey et al. (2011) and Munoz and Mackay (2019). Students requiring any timing exception were accommodated on a case-by-case basis. Based on LMS logs reporting the students' actual times spent completing the assessment and student feedback in synchronous online sessions, the authors found this approach met students' needs.

Other class activities were also modified. The term paper was broken into three separate components, and the instructor provided timely feedback on each part. This process provided constructive feedback quickly, both increasing student understanding and decreasing the chances a student could easily violate academic integrity using a paper mill (Rodchua, 2017). The group presentation was discarded, as the logistics of a group presentation in an online course were challenging with respect to time availability and cost/benefit outcomes. More emphasis was placed on the individual presentation, which could be a recorded presentation; students then asynchronously engaged in discussions moderated by the presenters. The final exam was configured identically to the first three midterm assessments. None of the assessments deployed any browser security.

Debriefing during the summer of 2020, the authors felt that the strength of the extant academic integrity safeguards was insufficient, an assumption shared by many educators across the country (The Wiley Network, 2020). This institution observed hundreds of students falling under suspicion of cheating in online exams when fewer than a dozen students had faced such suspicions in prior semesters. At the same time, it was clear that students were facing numerous stressors, which influenced the authors to build more class supports and academic safeguards.

In the fall of 2020, further changes were incorporated to support students and maintain academic integrity. Class progress checklists were added at the top of the LMS course to prompt students to install and familiarize themselves with the set of free apps and programs that they would need. All four assessments, including the final exam, incorporated randomization of multiple-choice questions. All assessments used a secure browser for administration; this locked student computers into kiosk mode, preventing the computer from accessing any resources other than the exam. A review of log files from the prior semester's assessments supported the adjustment of the time limit to 100 minutes. To reduce confusion and respect the remote nature of attendance, students were assigned specific presentation weeks and time blocks during which they were responsible for engaging in an online critique as part of the participation requirement.

Midway through the semester, as students struggled to stay current in their required activities, animated GIFs were programmed into the class LMS page to remind students of imminent due dates for upcoming activities. A graphic indicator of completion progress was also added to the class page; this completion taskbar was so impactful that 100% of students who were falling behind either caught up or reached out to the instructor within 24 hours of its appearance.

Course 2

In the spring of 2020, Course 2 began in person. Assessments included three open-book, take-home exams; a term paper; short individual presentations with ensuing discussions; a participation component based upon attendance and contribution to discussions; and a final exam which was also an open-book take-home assessment with three days allocated for completion. When the institution pivoted to distance learning, the discussions and class meetings incorporated synchronous online tools. No additional exam security was implemented, as the exams were already open resource in nature. This process was chosen as the best option for course continuity due to the technical nature of the class and the limited time available to develop a more synchronous approach.

Course 3

In the fall of 2020, Course 3, which had initially been designed as an asynchronous course, was retooled in accordance with College guidelines to include biweekly synchronous optional meetings. The term paper was changed from a single-submission document to a three-assignment project, animated GIFs were implemented as in Course 1, and students chose their presentation and critique weeks. This course involved more computational activities than the other courses, and the multiple-choice questions on all assessments were randomly selected from question banks, so no browser security measures were implemented.

Development of Secure Assessment Mechanism

In the spring of 2021, in response to on-campus concerns regarding academic integrity and cognizant of concerns raised by other research (Holden et al., 2021) and the need for some measure of accountability, the authors modified the assessment procedure in both Course 1 and Course 2 to include in-person virtual proctoring using Zoom. After the first assessment, students in both classes made it abundantly clear that the in-person proctoring via Zoom felt both invasive and logistically complex and increased their anxiety levels. The instructor asked them to try it one more time, thinking that perhaps the newness of the procedure was at fault. It was not; students even more vociferously protested after the second assessment, again contacting the instructor via email, text, the class LMS chat, and during class discussion time to voice their concerns. In response, the authors changed the proctoring mechanism to recordings captured by the institution's existing performance assessment tool; assessments still were administered within a secure browser.

Students were provided three weeks' notice regarding the new procedure, and the College provided spaces where students could take the exams if they did not wish their personal location to be recorded. Only the instructor and the LMS administrator had access to the recordings. The recordings were viewed only if separate indicators of potential academic integrity violations were observed. The recordings were maintained in accordance with institutional privacy policies.

By the end of this term, many faculty members had moved away from using Zoom as a proctoring tool. To safeguard academic integrity, these faculty members instead were relying on mechanisms such as more stringent question randomization selection, alternate types of questions, or alternative types of assessments completely. Twenty-two faculty members teaching courses incorporating various methods of online assessment, as well as three faculty members whose courses had only in-person assessments, agreed to engage students with a voluntary post-course survey of technology use in exams. Across the 25 courses, 77 students responded to an email request from their professor to take a Qualtrics survey following the college's Institutional Review Board procedures; the low return rate is attributed to the necessary timing of the survey, which was solicited after the final exam was completed. In this survey, students self-selected the assessment mechanism of their course and then responded to two Likert-scale prompts. Twenty-four students indicated their assessment had been a take-home assignment, 16 students' assessments were delivered via the LMS without security, and 15 students were unsure what mechanism had been used. Seven students were administered in-person assessments, while two students described their assessment as "secure"; the remaining 14 students were able to categorize their assessment more fully as either Zoom-proctored or recording-proctored. The responses are summarized in Table 1 (regarding perceived intrusiveness) and Table 2 (regarding perceived academic integrity guarantees).

Table 1

Student Perception of Intrusiveness of Assessment Mechanism (Spring 2021)

Assessment Mechanism	I felt the process was intrusive and/or invasive					
	Strongly disagree	*Somewhat disagree*	*Uncertain / Neutral*	*Somewhat agree*	*Strongly agree*	*Total*
Assignment	16	3	3	1	1	24
Moodle	9	2	2	3		16
Unsure	3	4	6	1	1	15
Record-proctor	2	3	4	2	1	12
In-person	4	1	2			7
Zoom-proctor	1			1		2
Secure				1		1
Total	35	13	17	9	3	77

Table 2

Student Faith in Academic Integrity Assurance of Assessment Mechanism (Spring 2021)

Assessment Mechanism	I believe this approach helped ensure . . . Academic Integrity					
	Strongly agree	Somewhat agree	Uncertain / Neutral	Somewhat disagree	Strongly disagree	Total
Assignment	15	8		1		24
Moodle	10	5	1			16
Unsure	1	2	2	4	6	15
Record-proctor	4	6	2			12
In-person	5	2				7
Zoom-proctor		1	1			2
Secure		1				1
Grand Total	35	25	6	5	6	77

The survey was targeted to courses that employed the full range of final exam administrations, such as in-person exams, take-home assignments, and unsecured and secured online exams. Data from the survey was combined with conversations from the authors' classrooms and asynchronous forums. This aspect was especially important given the limited number of responses from students who indicated their assessment had been Zoom-proctored; anecdotal comments from colleagues led to the conclusion that many of them, consistent with the authors, had switched away from Zoom-proctoring midway through the semester due to student complaints.

Survey results, combined with the student and colleague comments, led the authors to conclude the recording-proctored model used was relatively unintrusive when compared to the other methods of online-secure delivery (Zoom-proctoring, secure, and Moodle/LMS). This conclusion relied particularly upon the comments from students in their courses who had experienced both Zoom-proctoring and recording proctoring. Students for the most part preferred the second proctoring option. The authors also concluded that recording proctoring is viewed as likely to preserve academic integrity. Most importantly, this method appears to be relatively well balanced between the desired characteristics of non-intrusiveness and maintenance of exam security.

Therefore, in the fall of 2021, both Course 1 and Course 3 retained the recording-proctored online assessment method, although prior student comments were heeded to clarify directions and streamline sample assessments. Separate student comments regarding the unusual stressors of remote learning were addressed as well; on each exam, students were allowed to select any five questions to not answer (skip). Few students expressed frustration with the assessment procedures this semester.

In the spring of 2022, faculty members at the college implemented some of the assessment techniques developed throughout the prior three semesters. Many faculty members added progress bars to their own online courses, and randomized multiple-choice questions in online exams; additionally, at least two departments implemented the recorded-proctoring approach to online assessments.

Discussion

Towards the end of the spring 2021 semester, a student remarked to one of the authors that the recorded-proctoring method felt even less intrusive than an in-class exam, because "even if my professor watches me taking the test, it's not while I'm actually taking the test." Similar student comments, plus anecdotal observations by other faculty members, indicate that students feel reassured they are not presumed to be cheaters and appear to perceive the intrusiveness of this method to be no different from other methods. In both the fall 2021 and spring 2022 semesters, students stated that the assessment procedures were well-explained, and the survey results and follow-up discussions with students in the fall of 2021 revealed that students are more confident in the ability of this mechanism to safeguard academic integrity than other methods. An online assessment mechanism which centers both student comfort and academic integrity protection can achieve both and must be implemented to achieve equity and reliable assessment outcomes.

This is reassuring since this institution looks toward full participation in intercollegiate course-sharing consortiums in the coming years. To do so, the college will need to have confidence in its online assessment mechanisms. The past three years have demonstrated that faculty and administrators will accept long-term practice and procedural changes, even when such changes affect fundamental aspects of the institution. In the 2023 academic year, one of the institution's three Schools is planning to implement the recording-proctoring model as the only acceptable online assessment model. The institution's registrar has created a room prioritization system that privileges instructors who teach in-person classes that encourage a physical/virtual attendance mix policy.

The development of this model also makes it clear that implementing a secure method of delivering assessments is not, by itself, sufficient to secure academic integrity; technology tools can only support a culture of academic integrity. Transparency with students regarding the rationale and implementation of such tools demonstrates the centrality of students and enhances existing, mutually respectful academic cultures.

Future Considerations

Higher education institutions and residential liberal arts institutions will grapple with concerns regarding academic integrity while demonstrating respect for student perspectives continuingly as online course offerings expand. While these topics are broad, pervasive, and far-reaching, it is possible to address them in a cost-efficient manner. This requires careful consideration of technological controls for academic integrity, the unique environment of small liberal arts colleges, and tool-specific applications in a continuing academic atmosphere of change.

While technology alone cannot ensure academic integrity, a low-cost structure can sustain the necessary respect for students to promote an ongoing culture of academic integrity. It is important that institutions explore methods where technology used to enforce integrity rules also prioritizes respect for student autonomy and considers the impact on student anxiety (Conijn et al., 2022). This is especially recommended in relation to concepts of Universal Design for Learning, as technology tools can be particularly helpful in facilitating multiple ways of presenting material and multiple means of demonstrating mastery of concepts (Rogers-Shaw et al., 2018). Additionally, while technology tools can protect academic integrity in specific assessment situations, such as online testing, institutions must take care that the deployment of such tools does not damage the academic integrity relationship by creating a perception that no student is to be trusted.

Any academic integrity discussions must both contextualize and recognize the institution's culture. Campus culture at residential liberal arts colleges is different from larger institutions. The lived experiences of faculty members and administrators at smaller institutions may lead to an incomplete understanding of resource intensity and thus the cost of support resources can quickly overwhelm budgets. Liberal arts colleges may find that the time-on-technology required of faculty using online resources is at odds with extant practices which prioritize the face-to-face experience (Rust, 2019). Scenarios such as these must be considered whenever an institution finds it necessary to adopt online tools.

The culture of institutions, which emphasize in-person learning, also can inadvertently clash with the accommodations necessary to effectively use online learning tools. Institutions may not be prepared for the flexibility of scheduling that students expect in online and hybrid courses, and many students may still view online courses as "easy" courses (Baker et al., 2021), which can lead to increased academic integrity violations when it becomes clear such courses are, in fact, not "easy." Colleges and universities must explore student perceptions as well as prevailing institutional culture in concert with the administrative and faculty ethos, or an imbalance of respect and academic rigor may surface.

Finally, as institutions implement online assessment tools, some degree of standardization and familiarity, with respect to tool-specific considerations, must be part of the decision process. The performance assessment tool used at this institution does not provide a mechanism to pass the status of a recording in progress to the LMS; because the instructor needed to verify each student's recording status, students could not seamlessly begin their examination. Instead, they had to engage in a multi-step process that demanded their concentration, causing additional stress during the evaluation process. However, the use of this tool, which is common in more than two-thirds of the departments on campus, appears to have fostered student trust in the method overall. Institutions should take note: specific security-enhancing tools must be adopted across the institution, or the lack of standardization will frustrate students and potentially damage the culture of academic integrity. In essence, a seamless design for monitoring online class performance should integrate ease of use characteristics, academic integrity preservation, and data retrieval to support integrity concerns in a non-intrusive implementation.

Summary

The movement to online courses and delivery methods necessitated by the pandemic demonstrated that many universities and colleges were not adequately prepared for this transition, especially within such a short time frame. Still, it appears possible for any institution, regardless of its size or pandemic-prior focus, to leverage existing educational technology tools and find a balance between respecting students' autonomy and preserving academic integrity.

At the authors' small liberal-arts, residential college, faculty began to use existing tools to provide rigorous student assessment within the context of non-intrusiveness and integrity preservation, while also pursuing cost effectiveness, student acceptance, and faculty comfort. The process discussed in this example was molded to fit the institutional culture. The current implementation appears to present a successful path forward. An overarching assumption is that the process in operation will be subject to continual modification predicated upon technology changes, student tolerance, academic rigor, integrity maintenance, cost parameters, and administrative support. This process did, and can continue to, demonstrate that the cost of respecting students is surprisingly little.

References

Al-Freih, M. (2021). The impact of faculty experience with emergency remote teaching: An interpretive phenomenological study. *IAFOR Journal of Education, 9*(2), 7–23. https://doi.org/10.22492/ije.9.2.01

Bailey, A., Vaduganathan, N., Henry, T., & Laverdiere, R. (2018). *Making digital learning work*. Boston Consulting Group. https://edplus.asu.edu/sites/default/files/BCG-Making-Digital-Learning-Work-Apr-2018%20.pdf

Baker, D. A., Unni, R., Kerr-Sims, S., & Marquis, G. (2021, April). An examination of the factors leading to students' preferences and satisfaction with online courses. *International Journal for Business Education, 161*, 112-129.

Bozkurt, A., Jung, I., Xiao, J., Vladimirschi, V., Schuwer, R., Egorov, G.,. . . Paskevicius, M. (2020). A global outlook to the interruption of education due to COVID-19 pandemic: Navigating in a time of uncertainty and crisis. *Asian Journal of Distance Education, 15*(1), 1-126. https://doi.org/10.5281/zenodo.3878572

Bruff, D. (2011, February 28). *Why do students cheat?* Vanderbilt University. Retrieved May 10, 2022, from https://cft.vanderbilt.edu/2011/02/why-do-students-cheat/

Cluskey, G. R., Ehlen, C., & Raiborn, M. (2011, July). Thwarting online exam cheating without proctor supervision. *Journal of Academic and Business Ethics, 4*(1), 1-7.

Conijn, R., Kleingeld, A., Matzat, U., & Snijders, C. (2022). The fear of big brother: The potential negative side-effects of proctored exams. *Journal of Computer Assisted Learning*. https://doi.org/10.1111/jcal.12651

Cutri, R. M., & Mena, J. (2020). A critical reconceptualization of faculty readiness for online teaching. *Distance Education, 41*(3), 361–380. https://doi.org/10.1080/01587919.2020.1763167

Daffin Jr., L. W., & Jones, A. A. (2018). Comparing student performance on proctored and non-proctored exams in online psychology courses. *Online Learning, 22*(1). https://doi.org/10.24059/olj.v22i1.1079

Dyer, J. M., PettyJohn, H. C., & Saladin, S. (2020). Academic dishonesty and testing: How student beliefs and test settings impact decisions to cheat. *Journal of the National College Testing Association, 4*(1).

Evmenova, A. (2018). Preparing teachers to use Universal Design for Learning to support diverse learners. *Journal of Online Learning Research, 4*(2), 142-171.

Harton, H. C., Aladia, S., & Gordon, A. (2019). Faculty and student perceptions of cheating in online vs. traditional classes. *Online Journal of*

Distance Learning Administration, 22(4). Retrieved July 19, 2022, from https://ojdla.com/archive/winter224/hartonaladiagordon224.pdf

Haslam, C. R., Madsen, S., & Nielsen, J. A. (2020). The ISPIM Innovation Conference – Innovating in Times of Crisis. In *Event Proceedings: LUT Scientific and Expertise Publications.*

Holden, O. L., Norris, M. E., & Kuhlmeier, V. A. (2021). Academic integrity in online assessment: A research review. *Frontiers in Education, 6.* https://doi.org/10.3389/feduc.2021.639814

Johnson, N., Veletsianos, G., & Seaman, J. (2020). US faculty and administrators' experiences and approaches in the early weeks of the COVID-19 pandemic. *Online Learning, 24*(2), 6–21. https://doi.org/10.24059/olj.v24i2.2285

King, C. G., Guyette Jr, R. W., & Piotrowski, C. (2009). Online exams and cheating: An empirical analysis of business students' views. *Journal of Educators Online, 6*(1), 1-11. https://files.eric.ed.gov/fulltext/EJ904058.pdf

Kitahara, R., Westfall, F., & Mankelwicz, J. (2011, May). New, multi-faceted hybrid approaches to ensuring academic integrity. *Journal of Academic and Business Ethics, 3*(1).

Munoz, A., & Mackay, J. (2109). An online testing design choice typology towards cheating threat minimisation. *Journal of University Teaching and Learning Practice, 16*(3). https://doi.org/10.53761/1.16.3.5

Palmer, A., Pegrum, M., & Oakley, G. (2019). A wake-up call? Issues with plagiarism in transnational higher education." *Ethics & Behavior 29*(1), 23-50. https://doi.org/10.1080/10508422.2018.1466301

Poster, G. (2021, April 30). Lockdown browsers fail to create a culture of academic integrity. *The Retriever.* Retrieved April 8, 2022, from https://retriever.umbc.edu/2021/04/lockdown-browsers-fail-to-create-a-culture-of-academic-integrity-they-invade-student-privacy-and-harm-student-health/

Rodchua, S. (2017). Effective tools and strategies to promote academic integrity in e-learning. *International Journal of e-Education, e-Business, e-Management and e-Learning, 7*(3), 168–179. doi: 10.17706/ijeeee.2017.7.3.168-179

Rogers-Shaw, C., Carr-Chellman, D.J., & Choi, J. (2018). Universal Design for Learning: Guidelines for accessible online instruction. *Adult Learning, 29*(1). https://doi.org/10.1177/1045159517735530

Rust, J. (2019). Toward hybridity: The interplay of technology, pedagogy, and content across disciplines at a small liberal-arts college. *Journal of the*

Scholarship of Teaching and Learning, 19(2). https://doi.org/10.14434/josotl.v19i1.23585

Seife, D. A. & Maxwell, R. S. (2020). Cheating in online courses: Evidence from online proctoring. *Computers in Human Behavior Reports, 2*. https://doi.org/10.1016/j.chbr.2020.100033

Snyder, T. D., de Brey, C., & Dillow, S. A. (2019). *Digest of Education Statistics 2018* (NCES 2020-009). National Center for Education Statistics, Institute of Education Sciences, U.S. Department of Education, Washington, D.C.

Suryani, A. W., & Sugeng, B. (2019). Can you find it on the Web? Assessing university websites on academic integrity policy. *2019 International Conference on Electrical, Electronics and Information Engineering (ICEEIE) 6*, 309-313. https://doi.org/10.1109/ICEEIE47180.2019.8981405

The Editorial Board. (2021, March 18). Test proctoring software that films students is an invasion of privacy and presents problems with equity. *El Camino College The Union*. https://eccunion.com/opinion/editorials/2021/03/17/test-proctoring-software-that-films-students-is-an-invasion-of-privacy-and-presents-problems-with-equity/

The Wiley Network. (2020, July 22). *Is student cheating on the rise? How you can discourage it in your classroom*. The Wiley Network. https://www.wiley.com/en-us/network/education/instructors/teaching-strategies/is-student-cheating-on-the-rise-how-you-can-discourage-it-in-your-classroom

U.S. Department of Education, National Center for Education Statistics. (2021, January). *Digest of Education Statistics*. Retrieved April 12, 2022, from https://nces.ed.gov/programs/digest/d20/tables/dt20_311.15.asp

U.S. Department of Education, National Center for Education Statistics, Integrated Postsecondary Education Data System (IPEDS). (2022). *Percent of students enrolled in distance education courses, by state and distance education status of student: 2020* [Data Visualization Tool]. Retrieved April 12, 2022, from https://nces.ed.gov/ipeds/TrendGenerator/app/build-table/2/42?rid=6&cid=85

Venable, M. A. (2022). *2022 Online Education Trends Report*. BestColleges.com. https://www.bestcolleges.com/research/annual-trends-in-online-education/

Privacy in the Online Writing Center

James Hamby

Face-to-face writing centers have always dealt with issues of security. This is because students' personal information, papers (both graded and ungraded), and private conversations are all part of the daily business of writing centers. The need for discretion, diligence, and a knowledge of the Family Educational Rights and Privacy Act (FERPA) has long been a part of many centers' tutor training regimen. These issues are changed or amplified in online tutoring sessions, and the rapid switch to online-only tutoring that many centers experienced in Spring 2020 due to the COVID-19 pandemic presented numerous problems for administrators and tutors. Yet privacy concerns were not limited to students, as tutors and administrators suddenly found themselves more vulnerable as well; thus, administrators had to quickly adjust procedures to decide upon new best practices for online tutoring. At the Margaret H. Ordoubadian University Writing Center at Middle Tennessee State University, we were averaging around 2,500 sessions per semester before the pandemic, of which only around 400 were synchronous online live chat sessions in our scheduling software, WCONLINE. When we switched to all-online tutoring in March 2020, we added asynchronous document drop sessions to our services to better accommodate students during the pandemic. We made it through that spring and summer without seeing too much of a decline in our volume of appointments, and the sheer number of students we served created situations where privacy potentially could have been compromised. Some of these situations were foreseen, while others were unexpected. Many of the new privacy issues we encountered occurred during document drop appointments, but other problems also arose in communication between students and the writing center and in communication between staff members online. With each circumstance, we learned more about developing best practices for our online center moving forward.

Privacy in Writing Centers

Writing centers provide tutoring in written composition and rhetoric for college students in all disciplines at all levels, from First-Year Composition to doctoral dissertations. As such, writing centers are spaces in which the potential for a breach of privacy is very great. Student information is gathered and stored every time a writer makes an appointment (Parsons et al., 2021). Student papers, graded and ungraded, are out in the open, creating the potential that they may be seen by others without consent. Because students sometimes write about very personal issues, they and their tutors may discuss emotions, trauma, and other potentially sensitive topics in crowded environments and students may be uncomfortable with some of these conversations being overheard (Driscoll & Wells, 2020; Im et al., 2020; Perry, 2016). Sometimes professors contact writing centers wanting to know if their students have attended, what advice the tutor gave, and if the tutor detected any plagiarism (Conway, 1998). These situations occur in writing centers constantly, and many more unexpected ones may arise. This is why every year we have a representative from the university counsel office give us a training session during our orientation on what FERPA is and what we should do to be aware of student privacy. Bridgewater et al. (2019) note that FERPA training is an essential part of becoming a writing tutor. Protecting student privacy not only fulfills the law, but it indicates to the staff the importance our writing center places upon students as individuals who are placing a great deal of trust in writing center associates.

Asynchronous Tutoring

The sudden shift to all-online tutoring in March 2020 created several new privacy concerns to which we had to quickly adapt. Most of these revolved around our asynchronous document-drop tutoring sessions, which were completely new to us. The most pressing problem was what to do about student papers downloaded to tutors' personal computers. In order to conduct an asynchronous session, students upload their papers to our online platform, WCONLINE, a popular website to which many writing centers subscribe. WCONLINE allows students to schedule their own appointments after signing up for an account, and it also allows writing center administrators to keep accurate records about how students utilize their writing centers. WCONLINE supports both synchronous live-chat and asynchronous document-drop tutoring. The synchronous live chats function much like a Zoom meeting, while in document drop sessions, students upload their papers and tutors then download them, make comments, and upload the new files back onto WCONLINE for the students to access. The document drop process left each tutor with a large cache of student work. As Parsons et al. (2021) note, the storing of writers' personal information may have a "chilling effect" on "intellectual pursuits" that would discourage students from coming back to the writing center (15). Since tutors were working on document drops from home, it was deemed unavoidable that they would be downloading papers to their personal devices. We encouraged them to delete student papers frequently, and we made a rule that required them to delete papers no less than once a week. Of course, this was an unenforceable rule, as there was no way we could know what they were doing on their personal computers. However, we did talk about deleting student papers frequently in our staff meetings and in our interactions over Discord (our use of Discord to communicate during shifts will be discussed below) and stressing the importance of student privacy helped set the expectation that we would keep student information confidential as much as possible.

Another challenge with privacy during document drops was protecting the identities of our tutors, as online tutoring created new situations that may violate their privacy or sense of safety (Nadler 2019; Prebel 2015). When making comments in Microsoft Word, the default setting for the name at the top of the margins comment box is usually the full name of the person to whom the Microsoft Word account is registered. While WCONLINE lists tutors' first names, we caution our tutors to not reveal their last names, their contact information, or any other information they would feel uncomfortable about somebody else having. Fortunately, this problem was easy to fix as the name on the Microsoft Word comment function can be changed.

The biggest impediment to doing this is that the steps are slightly different in each version of Word. For the most part, these steps can be found in a Google search, but some tutors still required help doing this. We found that we needed to check in with each tutor because asking for technical help can be embarrassing, especially when it seems that one's colleagues are having no difficulty with performing the same task. This emphasized to us administrators that we should not take it for granted that all tutors (or students, for that matter) are at the same level of comfort with technology, and that whenever we add a technology or require a new procedure, we need to provide adequate training.

Online Synchronous Tutoring

Our live chat appointments also presented a new set of challenges—but, fortunately, we had implemented audio/visual synchronous tutoring in Spring 2018. When we first instituted audio/visual tutoring, we were surprised by the reluctance of some of the staff. Such resistance to change, however, is not uncommon for writing centers (Neaderhiser & Wolfe, 2009). Though tutors were perfectly fine with meeting students in person for face-to-face sessions, they balked at being on screen for live chat sessions, because the students would be able to see them. However, most of these anxieties faded over time as they became more comfortable with live chat tutoring. Tutors new to the center the next semester accepted live chat tutoring as a part of the job, so it seemed to us that the initial resistance to live chat tutoring was largely due to its novelty and not, as some tutors suggested, to any uncomfortable social circumstances between tutor and student.

With the switch to all-online tutoring in Spring 2020, we feared a similar situation would happen. However, at that time, so much of education, business, and everything else was moving to Zoom and similar platforms, that shifting tutoring to all-online seems to have been accepted by our staff as a necessity. While they may have been more comfortable with face-to-face tutoring, they were all willing to adapt their practices for live chat sessions at home. Moreover, our administration and tutors embraced the opportunity the pandemic presented in using new methods to make our center more technologically flexible, which has been a major movement in writing centers now for decades (Andersen & Molloy, 2022), which also served to bolster our center's abilities for access and inclusion (Bell et al., 2022).

What was new in this situation, however, was that now students and tutors were regularly seeing into one another's homes. Tutors, of course, had been able to see into students' homes before this, but not as frequently. In previous semesters, students often participated in live chat appointments from the library, from cafes, or from secure spaces in their homes. They had the freedom to choose where they participated in their online sessions and whether they wanted to reveal anything about their home lives. During the early days of the pandemic with mandatory quarantines, however, students lost this autonomy and were often attending tutoring sessions in houses filled with family members or roommates, leaving little room for privacy.

Inevitably, some of the tutoring sessions were overheard by others in the students' homes, but what we most worried about were helicopter parents hovering around their student and wanting to listen in on the session. We have had this situation happen before in the physical center, but we always politely told the parents that all consultations were private and that they would have to wait outside. However, with students attending sessions in their homes, the potential for parents listening was much greater. This was compounded with the problem that we would not necessarily know if the parents were listening in, as they could easily just stay out of the camera's scope. Fortunately, to our knowledge, this situation never arose, but we did talk to our staff about strategies for addressing this circumstance and why it was important for consultations to remain confidential.

While we did not have parents listening in to sessions, we worried about violating students' privacy in other ways. Having a stranger, especially a university employee, seeing into their home can be very disconcerting for college students. H. Denny and Towle (2017) observe that tutoring sessions often embrace "the crosscurrents of wider social, economic and cultural relations" (para. 4). A student may feel judged if their room is cluttered or if there is something else going on in the background, such as additional family members being overheard in the background.

Many students who were caring for children often became dismayed when the children wanted to interact with them, were crying, or were simply playing loudly. Our center has always been pro-family, and it is not uncommon for students, tutors, and administrators alike to bring their children with them into the center. As with other situations, we talked with our staff about tutoring students with children, and we stressed that empathy and understanding were important during these times. And though our staff responded positively to these sentiments, many students still chose to end their sessions early when they felt their home situation was untenable. Perhaps these students were worried that they would be judged and thought not dedicated to their studies. Manze et al. (2021) report that students who are also parents that participated in their study on the effects of the pandemic on college classrooms often felt that "they were hesitant to ask for accommodations, not wanting to be perceived as opportunistic or manipulative" (p. 635). In the same way that students' family members created potential privacy issues, so too did tutors' household members. We asked our tutors to be mindful of who was around while they were conducting sessions so that nobody would overhear their discussions with tutees as the subject matter of student papers sometimes requires more privacy.

As Prebel (2015) observes, "Writing Center work frequently involves a willingness to talk about the self and deeply personal experiences, including trauma" (p 2-3). Students often take advantage of writing assignments to process emotions they are experiencing; for many, this may even be the first time they confront these feelings. The prospect of having to talk about emotions or traumatic experiences within earshot of either the student's own family or roommates or of the tutor's household members could have a negative effect on the opportunity for frank and honest conversation. Even if a tutoring session does not involve something as serious as discussions of trauma, other factors may make a student feel trepidatious about being overheard by a tutor's housemates. As H.C. Denny (2010) notes, "People's access to education and literacy is charged with politics and carries the weight of wider historical relations, all of which impact on their sense of agency and facility with writing for particular discourse communities, most often the academic" (p. 88). It is often intimidating enough for students to overcome whatever barriers they may feel in coming to the writing center and sharing their writing with tutors in the first place, but to further risk being overheard by others may exacerbate those fears.

Tutors, too, may have reasons they do not want students to be aware of their family members. As Tondy et al. (2022) note, tutors in online settings at home lose a great deal of control as they have to "function professionally within home environments wrought with noise, family, pets, and other distractions" (para. 20). Of course, some of our tutors were juggling childcare responsibilities with their shifts, and at times kids would make an appearance during sessions. We again took an approach of empathy, but no one is ever comfortable working and parenting simultaneously. Less scrupulous employers (even in writing centers), may hold parenting responsibilities against workers, thus creating a breach of employee privacy. Even less drastic breaches can be an invasion of an employee's privacy as their home becomes their workplace. An untidy room, silly posters on the wall, or an outdated computer not functioning properly could cause a tutor to worry that they are at risk of being judged as unprofessional, but it is not fair to judge a tutor's professionalism by their home environs, especially when they are under quarantine during a global pandemic. Claman et al. (2021) argue that tutors "taking appointments from the comfort of their own home ignores the way that bringing the many spheres of life into their lodgings disrupts the constancy that they wish for when occupying their home" (para. 7). Even if there is nothing wrong with a tutor's house, it can still be disconcerting for them to allow strangers to see into their homes.

Communication

One more potential breach of employee privacy affected the administrators directly. Although the entire university had switched to online operations and the writing center's phone message explicitly stated that we were not answering the phone and to please simply email us, several students still left voicemails. While we were able to access these voicemails from home, we were unable to call back except from our personal phones. Of course, this situation was not ideal, and it did result in students mistakenly thinking our numbers were from office phones and calling us back at awkward times. However, this was at least a decision that we made for ourselves, and we stressed to our tutors that we did not want them, under any circumstances, to use their own phones to call back students who had left a message.

A tutoring staff that works entirely from their homes can also lead to an overly relaxed atmosphere where tutors forget to be confidential about discussing sessions. Before the pandemic, when a vast majority of our sessions were in the physical center, we always discussed with our staff the importance of keeping all conversations professional in tone and content, including not discussing tutoring sessions. When we moved to all-online tutoring, one of our tutors recommended adopting Discord, a social media platform originally designed for video games, as a means of communication during tutoring hours. Discord worked wonderfully, both for conducting the business of the center as well as for providing a social outlet for our quarantined staff, many of whom keenly wanted social interaction (Carter et al., 2020). However, the relaxed nature of communicating over Discord perhaps made it too easy for tutors to be forgetful about professional standards. We realized this when a tutor mentioned a student by name and said that they were having a problem with a specific aspect of their writing. While we encouraged tutors to ask questions of one another over Discord during sessions, such as "Does anyone have a good online source for integrating quotes?", we felt referencing a student by name was a privacy violation. This led to us training the staff on appropriate ways to use Discord that respected students' anonymity. We told them that if they needed to pose a question to the group chat, they should make it general enough to prevent anyone from identifying that question with a particular tutoring session. If the question did need to be specific to a particular student, then they should directly message an administrator.

We also experienced an uptick in professors wanting proof of attendance for students who had attended the writing center. Fortunately, we already had a good system in place for this event. For many years now we have recognized that discussing students' writing center appointments with their professors, or even acknowledging that they attended, can be a privacy violation (Conway, 1998). Whenever a professor contacts us about a student, we explain this policy to them, and we encourage them to not require proof of attendance as a part of their course, both for privacy reasons and to prevent our center from being overrun. However, we do let students and professors know that, at the student's discretion, they may share their client report form (CRF) with the professor as proof of attendance. Additionally, the student may share the whole CRF with their professor or just the email that shows they have received one after an appointment. This practice gives the student agency in determining how much information they wish to share, if at all.

Student Trust

In addition to the difficulties of securing privacy in online writing centers, it is also difficult to create a sense of rapport with students in virtual settings. Hewett (2015) notes that the ability to establish "a trusting relationship with students in online settings involves a wide variety of activities that both build relationships and solve problems" (p. 48). Writing center pedagogy relies upon the ability to quickly build rapport with students. Sharing writing with a stranger and asking them to help requires an enormous amount of trust on the side of the student. Writing is a very personal endeavor, and students need to feel comfortable in writing center environments, whether they be face-to-face or online, and it is much more difficult to build rapport in an online environment. Youde (2020) notes that emotionally intelligent tutors "can create a more open and effective learning environment with fewer distractions" (p. 25). However, if a student feels their privacy is not secure in an online writing center, then they will be reluctant to use that service again.

Best Practices

Maintaining privacy in online writing centers can be challenging, but we have established best practices over the past two years that have helped us, and we have incorporated these practices into our orientations and weekly staff meetings. Firstly, in privacy concerns with technology, center directors should always be aware of how information is stored. This pertains to both records in online platforms and to any student information that tutors may download on their computers at home. Administrators should have strict guidelines about deleting sensitive information once it is no longer needed. It would be best for at-home tutors to be assigned university-owned laptops so that the university can ensure all stored information is deleted. However, this is a major budgetary issue and may not be feasible.

Communication between tutors and students and between tutors and other tutors also presents privacy challenges. When responding to student papers through Microsoft Word comments, tutors should always change the settings so that their last names are not revealed. Tutors should be aware of the statements that they make over social media, email, scheduling platforms, and whatever other forms of electronic communication their online writing center utilizes. When tutoring from home, employees should also be encouraged to isolate themselves as much as possible in order to protect the student's privacy.

Unfortunately, there is very little tutors can do to ensure that students' family members are not listening in on their sessions, but administrators should discuss this issue with their tutors so that they are aware. Merely asking that students be alone for their sessions may be enough to deter a well-meaning parent who did not think of listening in as a privacy violation. Finally, writing centers should determine what their policies are for professors who require proof of attendance as a part of their course or for extra credit. The circumstances of these requests and how centers respond may vary from institution to institution, but administrators should come up with some system that is clear, consistent, and that keeps the privacy of students at the forefront.

Conclusion

Much of higher education was already trending towards online options before COVID-19, and the rapid switch to online learning at the onset of the pandemic necessitated and accelerated those trends. For many students, support services like online writing centers have been a vital component of their education over the past two years. Hopefully, students' participation in online centers has disrupted the notion that remote learning is a solitary pursuit. Additionally, what little human interaction students experienced during their online sessions may have given them a sense of community during a very lonely time. However, the shift to online tutoring created many potential privacy breaches and writing center administrators will have to continuously evaluate privacy procedures as technology continues to evolve. Protecting students' privacy is not only the right thing to do, but it is an essential part of creating a welcoming, safe community that facilitates student learning.

References

Andersen, E. M., & Molloy, S. (2022). Retooling the OWC: Offering clients online platform choices during a pandemic. *WLN: A Journal of Writing*

Center Scholarship, 46(9-10), 3-10.
https://www.wlnjournal.org/archives/v46/46.9-10.pdf

Bell, L. E., Brantley, A., & Van Vleet, M. (2022). Why writers choose asynchronous online tutoring: Issues of access and inclusion. *WLN: A Journal of Writing Center Scholarship, 46*(5-6), 3-10. https://www.wlnjournal.org/archives/v46/46.5-6.pdf

Bridgewater, B., Pounds, E., & Morley, A. (2019). Designing a writing tutor-led plagiarism intervention program. *Learning Assistance Review, 24*(2), 11-27. https://files.eric.ed.gov/fulltext/EJ1234299.pdf

Carter, K., Cirillo-McCarthy, E., & Hamby, J. (2020, August 10). Creating harmony through Discord. *Connecting Writing Centers Across Borders: A Blog of WLN: A Journal of Writing Center Scholarship.* https://www.wlnjournal.org/blog/2020/08/creating-harmony-through-discord/

Claman, A., Seekins, C., & Mardell, S. (2021). Sheltering in place, working in space: Reflections on an online writing center. *The Peer Review, 5*(2). https://thepeerreview-iwca.org/issues/issue-5-2/sheltering-in-place-working-in-space-reflections-on-an-online-writing-center-at-home/

Conway, G. (1998). Reporting writing center sessions to faculty: Pedagogical and ethical considerations. *The Writing Lab Newsletter, 22*(8), 9-12.

Denny, H. C. (2010). *Facing the center: Toward an identity politics of one-to-one mentoring.* Utah State University Press.

Denny, H. & Towle, B. (2017). Braving the waters of class: Performance, intersectionality, and the policing of working class identity in everyday writing centers. *The Peer Review, 1*(2). https://thepeerreview-iwca.org/issues/braver-spaces/braving-the-waters-of-class-performance-intersectionality-and-the-policing-of-working-class-identity-in-everyday-writing-centers/

Driscoll, D. L. & Wells, J. (2020). Tutoring the whole person: Supporting emotional development in writers and tutors. *Praxis: A Writing Center Journal, 17*(3), 16-28. https://repositories.lib.utexas.edu/bitstream/handle/2152/82567/396_Driscoll_Wells_Proof_8_10.pdf?sequence=2

Hewett, B. L. (2015). *The online writing conference: A guide for teachers and tutors.* Bedford/St. Martin's.

Im, H., Shao, J., & Chen, C. (2020). The emotional sponge: Perceived reasons for emotionally laborious sessions and coping strategies of peer writing tutors. *Writing Center Journal, 38*(1-2), (203-28). https://www.jstor.org/stable/27031268

Manze, M. G., Rauh, L., Smith-Faust, P., & Watnick, D. (2021). Experiences of college students with children during the COVID-19 Pandemic. *Emerging Adulthood, 9*(5), 631- 38. https://doi.org/10.1177/21676968211020225

Nadler, R. (2019). Sexual harassment, dirty underwear, and coffee bar hipsters: Welcome to the virtual writing center. *The Peer Review, 3*(1). https://thepeerreview-iwca.org/issues/redefining-welcome/sexual-harassment-dirty-underwear-and-coffee-bar-hipsters-welcome-to-the-virtual-writing-center/

Niederhiser, S. & Wolfe, J. (2009). Between technological endorsement and resistance: The state of online writing centers. *The Writing Center Journal, 29*(1), 49-77. https://doi.org/10.7771/2832-9414.1670

Parsons, M., Dolinger, E., & Tirabassi, K. E. (2021). Good to know? Confidentiality and privacy in writing centers and libraries. *WLN: A Journal of Writing Center Scholarship, 45*(9-10), 10–17. https://www.wlnjournal.org/archives/v45/45.9-10.pdf

Perry, A. (2016). Training for triggers: Helping writing center consultants navigate emotional sessions. *Composition Forum, 34*. https://compositionforum.com/issue/34/training-triggers.php

Prebel, J. (2015). Confessions in the writing center: Constructionist approaches in the era of mandatory reporting. *WLN: A Journal of Writing Center Scholarship, 40*(3-4), 2-7. https://wlnjournal.org/archives/v40/40.3-4.pdf

Tondy, E., Gelet, A., & Wetzl, A. (2022). Tutor metamorphosis: Expectations and reality when tutoring remotely. *The Peer Review, 6*(1). https://thepeerreview-iwca.org/issues/issue-6-1/tutor-metamorphosis-expectations-and-reality-when-tutoring-remotely/

Youde, Andrew. (2020). *The emotionally intelligent online tutor: Effective tutoring in blended and distance learning environments*. Routledge.

Artificial Intelligence for Privacy Conservation in Remote Learning

Hongbo Zhang, Lei Miao, Jia-Xing Zhong, and Aimin Yan

COVID-19 resulted in a significant impact on academic learning when schools shut down across the world. Globally, over 1.2 billion children were out of the classroom during the pandemic (Li & Lalani, 2020). Remote learning does offer an alternative learning mode when in-person instruction is infeasible; however, learning outcomes of remote learning are mixed. The effectiveness of remote learning has been compromised due to the lack of in-person interactions between students and instructors. In addition, student privacy is difficult to maintain in front of a camera. This leads to hesitation to show the student's complete profile, including body language and environmental contexts, which therefore reduces the engagement, effusiveness, and immersion of the overall learning process (Yang et al., 2020).

There are numerous methods available for preserving user privacy during remote video calls. Among them, a change of background is often used. Through revision of the background, students can disable their actual background which can contain sensitive and private information. Changing a remote video call's background is made feasible through several different techniques. When replacing the actual background with a virtual background, image matting is one major technique used for this task. This chapter will review different image matting techniques for how image matting is implemented.

For preserving privacy, the suppression of background noise is also important. Background noise is one primary source of privacy concerns. Suppressing background noise and only exposing the student's intended sound to others will improve privacy. Background noise removal is different from complete noise removal. As such, simple noise removal will not work. This chapter will discuss different techniques for recognition of the speaker's sound features and suppression of others. Different techniques involving recurrent neural network, conditional GAN network, and WaveNet will be discussed.

The blurring of shared screen content dynamically and selectively is also an important technique for preserving privacy. For example, a student or teacher may not want other people to visualize the websites that they have visited, their desktop background, or the apps they have downloaded or been using. The current method for achieving this is through a selective sharing mechanism, where students or the teacher share a particular window. However, it can be inconvenient for students to share only a particular window since they may need to switch to different apps dynamically through the process. Therefore, it will be helpful for computers to smartly blur the privacy related content while still showing other content to participants. For this, deep learning-based natural language processing models will be reviewed for how to recognize content specific to the meeting while blurring others.

Methods

Virtual Background for Privacy

Overview of Virtual Background for Privacy
Image matting is one primary technique used for replacing the actual background with a virtual background during remote meetings. Image matting is the process of estimating foreground objects in images and videos. The process starts by estimating the dimensions of the foreground images and then extracting the foreground from the background. In cases of transparent objects, the situation would become complex since the pixels would belong to the foreground and background at the same time. Such objects include human hair and animal fur, which require estimating the transparency values of the object.

The major differences between image matting versus image segmentation is the introduction of the alpha channel. When the alpha value is one, it is pure foreground. Conversely, when the alpha value is zero, it is pure background. When the alpha value is between zero and one, it is part of both the foreground and background, therefore it is a mixed value. However, there are limitations of such a matting-based method. The method is based on the color differences for differentiating the foreground and background. Color can be very different based on different lighting and environmental conditions, even for the same object, hence it is not reliable. Second, the generation of the ground truth for matting is known to be very difficult. It involves labor extensive human-image interaction work for generation of the ground truth. Due to this, the current available datasets are small. Most of them contain small ground truth (around 100 or 1000 images or videos). This further makes the training of image matting quite challenging (Xu et al., 2017).

Methods of Virtual Background for Privacy

Different methods are used for image matting, including sampling-based methods such as ray casting, searching the entire boundary, and sampling from color clusters (Feng et al., 2016; Gastal & Oliveira, 2010). Another method also includes the measurement of the distance of samples from a known pixel to conclude the similarity of the foreground and background (Levin et al., 2007). Similarly, KL divergence and sparse coding approaches are also used for such sampling-based methods (Bu et al., 2018). The average of the foreground and background are assumed to calculate the foreground, background, and alpha value (Levin et al., 2007). More statistically meaningful research has modeled the foreground and background using Gaussian distribution and therefore uses statistical learning methods for image matting (Chuang et al., 2001). Likewise, the matting problem can also be formulated as a Poisson equation form between the foreground and background. The Poisson equation models the matting gradient field and Dirichlet boundary conditions of the foreground and background, hence numerical methods are used to solve the Poisson equation (Sun et al., 2004). Other approaches attempt to formulate the matting problem as an optimization problem. The approach assigns a regularization term to a pixel and then optimizes the belonging of the pixel to either the foreground, background, or both (Levin et al., 2007). Furthermore, deep learning methods have gained ground in image matting. Different methods rely on different mechanisms for image matting. One approach is to learn the TriMap method through the matting process.

GAN networks have also started advancing to generate virtual backgrounds, hence GAN techniques for generation of virtual background will also be discussed. Among them, conditional Generative Adversarial Networks (cGAN) have been successfully used for background subtraction. Within this method, the input for the generator is the image and background, and the output is the foreground mask. The discriminator learns to differentiate the real from the fake foreground. Studies show that with CDnet 2014 and BMC databases, the proposed cGAN method can achieve appreciated performance for background subtraction (Bakkay et al., 2018). Other cGAN methods seek a different training pipeline for background removal. In this method, the generator is trained to generate images without background and the discriminator is trained to remove background, which is different from traditional background removal where an image-to-semantics process is taken (Wang et al., 2020). For removal of dynamic objects from the scene, a more dynamic approach such as moving object segmentation should be used. With this method, the generator produces dynamic backgrounds similar to the test sequences to increase the generation fidelity. In particular, the training considers the loss function in image space and feature space such that the generated images have superior performance in both these two spaces (Sultana et al., 2022). Compared to convolutional neural network (CNN) methods, the GAN network method is appealing since it can generate the foreground image instead of being primarily limited to segmentation as compared to CNN methods. Free of such limitations, the method is promising for remote learning where obtaining large scale semantic segmentation is a challenge.

For practical background removal for remote learning, real time background removal is critical. Real time background removal requires a simple computation pipeline. For this end, numerous methods have been proposed. Among them, a TriMap-free method requiring less annotation effort is appealing (Sengupta et al., 2020). This method asks the user to first show the background image without users in the scene. It is followed by the generation of the alpha matte with the deep network, along with input of the pre-captured background image. Then, an image of the user is created in the scene as a soft segmentation image, and where motion cues will be considered. Specific consideration of motion cues will make the method practical for remote learning, where the scene is dynamic rather than static. To further improve the fidelity of the background subtraction for immersive remote learning, a self-supervised GAN training pipeline is used. In comparison to a non-GAN-based training pipeline, the discriminator does not need labeled images during the training phase, therefore it is feasible to train on large scale customized data.

Another practical concern of the privacy conservation for remote learning is the dynamic scene of the learning environment. As such, methods for removal of background through dynamic scenes are critical. Moving object segmentation based dynamic background subtraction is therefore proposed for this purpose (Patil et al., 2021). Within this method, multiple frames of the remote learning video are explicitly considered. Hence, the motion cues are built through the consecutive frames of the video. The motion cues are useful in tracking highly dynamic motion objects to achieve ideal background subtraction. Furthermore, the dilated convolution is used to extract multiple scale features from the scene to achieve better feature extraction for learning foreground and background. Then a GAN based generator and discriminator are used to further enhance the background subtraction quality to reduce the subtraction artifact. Results show that the method can remove background objects from dynamic scenes such as walking humans and driving cars. This method is therefore potentially useful for outdoor-based learning environment privacy conservation.

Challenges and Problems of Virtual Background for Privacy

The current background subtraction methods still present a few challenges. Primarily, no background subtraction methods have considered the removal of unexpected objects, such as a child suddenly appearing in the remote learning scene. The current background subtraction methods have only considered the static scene. As such, the abrupt appearance of a child in the remote learning scene is a challenge for the background subtraction to remove. Methods able to effectively track dynamic scenes are better at dealing with this abrupt appearance of an object (Patil et al., 2021). However, the current dynamic scene background subtraction methods have mostly only considered the tracking of foreground objects rather than the background objects; hence, modification of the existing methods is necessary. The second major challenge is the use consideration of the outdoor environment. At the present, most methods are designed to work for static scenes such as an indoor environment. For outdoor environments, the scene is rather complex. Frequently, moving objects are present in the background, and most existing methods are not ready to handle this condition. The last major challenge is to selectively replace background elements. Often, learners will not want to replace the whole background scene, but the existing methods mostly do not consider this. Therefore, selective background subtraction is difficult to achieve at the present.

Specific to remote learning, students are required to turn on the video camera to expose their test environment for teachers to monitor cheating during test-taking. The desired outcome is for students to take tests without cheating; however, student privacy is invaded through the process. Under this condition, selective exposure of the student's background is critical to ensure teachers can visualize the student's environment while also protecting their privacy. However, in practice, it is extremely challenging to achieve this balanced goal. It is difficult to know which objects are privacy-sensitive in the student's environment. It is also possible that what students believe should be private may actually allow students to cheat. In order to achieve such a balanced goal, a large-scale database needs to be built to understand privacy-critical objects and the objects that allow students to cheat through the remote test. Similarly, once the database is built, deep learning-based methods could be used to train a background subtraction model for selective background subtraction for ideal privacy conservation.

It is also common that teachers may record the remote learning sessions. Through the recording, information related to the student's environment persists in either a local or remote database. Most often, the database is stored in a cloud learning platform such as Blackboard or Canvas. The recorded remote learning session is available to be viewed by all other students. A practical concern is that the recorded video might contain private information. Therefore, it is desirable to remove such *privacy-sensitive* background information. However, until present, it is largely an open research question as to how to define the *privacy-sensitive* information and consequently to remove it. Fortunately, there are a few video background removal models for us to use, which includes a real time video background subtraction method, the segmentation method (Cioppa et al., 2020). Within the segmentation method, a real time object classifier is introduced through consideration of inter frame information cues. It can achieve better semantic segmentation of the foreground objects. In order to achieve real time subtraction of the video background, a simple Manhattan distance (the distance between two points in a grid (Black, 2019)) between the current pixel's color and ground truth is used for making the alpha matting map. Such a simple threshold-based rule enables real time background subtraction. Video background removal is limited to videos seen by computers, which are restricted by the number and types of training videos. For the general-purpose type of videos, an unseen video background removal is proposed (Tezcan et al., 2021). The method takes both spatial and temporal information of the video into consideration and uses data augmentation such as spatial-temporal crop and spatially aligned crop techniques to generalize the types of videos for background subtraction. It is expected that the video background removal will enable long-term privacy conservation for remote learning, thus easing students' and parents' concerns about participating and engaging in remote learning.

Suppression of Background Sound for Privacy

As one of the main potential causes of privacy disclosure, background noise leakage in online calls is increasingly emphasized by both parties in the remote learning process. For the purpose of background noise suppression, speech enhancement (SE) technology is adopted ubiquitously to online meeting software (such as Zoom, Teams, and Skype). For this, the chapter will describe the status quo of AI technology in background noise suppression from the following two perspectives: methods and their performances.

Methods of Suppression of Background Sound

The existing methods can be roughly categorized into two groups: traditional statistical models and deep learning models. The statistical models usually hypothesize that the noisy observations are based on stationary background noises, which makes it highly difficult to deal with real-world scenarios with non-stationary noises. Owing to the strong modeling capacity of deep neural networks (DNNs), it is feasible to apply deep learning to background sound suppression in the non-stationary setting.

Traditional Statistical Models. Ephraim and Malah (1985) proposed a short-time spectral amplitude (STSA) estimator and examined it while enhancing noisy speech. By finding the minimum of the log-spectral mean square error between the original STSA in the speech signal and its estimate, this estimator showed effectiveness in improving the quality of noisy speech. To provide simpler alternatives to the STSA rule, Wolfe and Godsill (2001) presented the Bayesian approaches. Under the same modeling assumptions, these approaches exhibited almost identical behavior to STSA. Compared with the unmodified STSA, they were efficient to implement and yielded intuitive interpretation. Lotter and Vary (2005) devised two spectral amplitude estimators for acoustical background sound suppression. The two estimators were based on the maximum a posteriori estimation and a super-Gaussian statistical model, respectively. These estimators were able to optimally fit the distribution of the speech spectrum for a background sound reduction system. Srinivasan et al. (2007) trained codebooks of speech and noise linear predictive coefficients. Furthermore, they developed both memoryless and memory-based estimators to obtain the minimum mean squared error estimate of the clean speech signal. In this manner, their proposed scheme performed well in a noisy background. For single microphone SE, Reddy et al. (2017) derived a gain function based on super-Gaussian joint maximum a posterior (SGJMAP). In the SGJMAP-based function, a tradeoff parameter is further introduced to customize the listening preference. Experimental results reflected the usefulness of this SGJMAP-based application in real-world noisy backgrounds.

Deep Learning Models. In contrast to conventional researchers on background noise suppression who focus on reducing the minimum mean square error (MMSE), Xu et al. (2014) attempted to find a mapping function between noisy and clean speech signals based on DNNs. Xu et al. regarded SE as a supervised learning task, in which case clean speech is provided as the fitting target on training datasets. This supervised learning paradigm has been adopted by many follow-up works as illustrated hereunder. Luo and Mesgarani (2019) developed a fully convolutional time-domain audio separation network (Conv-TasNet) for end-to-end time-domain speech separation. To separate individual speakers, Conv-TasNet encodes a representation of the speech waveform spectrum and inverts it back to the waveforms via a linear decoder. Likewise, Pandey and Wang (2019) put forward another fully convolutional neural network for real-time SE, which is dubbed a Temporal Convolutional Neural Network (TCNN). Under the supervision in a speaker- and noise-independent way, TCNN encodes a low-dimensional representation of a noisy input frame and decodes the representation to reconstruct clean speech. In addition to amplitude prediction, Yin et al. (2020) address the problem of phase prediction by putting forward a Phase-and-Harmonics-Aware Speech Enhancement Network (PHASEN). As an innovative framework, PHASEN captures long-range correlations along the frequency axis and does well in time-frequency spectrogram reconstruction. Ephrat et al. (2018) introduce a joint audio-visual model to separate a single speech signal from a mixture of audio such as other speakers' voices and background sound. This method shows superiority in audio-only speech separation in cases of mixed speech, and it is a speaker-independent solution (trained once, applicable to any speaker). To enable isolated control over the importance of speech distortion versus noise reduction, Xia et al. (2020) devise two mean–squared error-based loss functions as the learning objectives. By optimizing these two objectives, the model achieves high performance in real-time single-channel speech enhancement. Koyama et al. (2020) propose a STFT-based method and a loss function with problem-agnostic speech encoder (PASE) features. By doing this, their model achieves excellent performance in the task of deep noise suppression. Westhausen and Meyer (2020) combine a short-time Fourier transform (STFT) with a learned analysis and synthesis basis in a stacked-network approach. By training a dual-signal transformation network on 500-hour noisy speech, the STFT-based method can suppress real-time background noise. Recently, Watcharasupat et al. (2022) have exploited the offset-compensating property of complex time-frequency masks and presented an end-to-end complex-valued neural network architecture. The presented architecture further utilizes a dual-mask technique, thereby simultaneously suppressing background sound and canceling acoustic echo.

Performance of Background Suppression

To evaluate the performance of noise suppression, researchers have proposed both objective and subjective metrics. The former aims to consider the sound quality not influenced by personal feelings, while the latter intends to correlate well with the testing results of human subjectivity.

Objective Metrics. There exist quite a few objective measures, e.g., Speech to Distortion Ratio (SDR) (Nocerino et al., 1985), Signal to Noise Ratio (SNR) (Johnson, 2006), Perceptual Objective Listening Quality Analysis (POLQA) (Beerends et al., 2013), Perceptual Evaluation of Speech Quality (PESQ) (Rix et al., 2001), and Virtual Speech Quality Objective Listener (ViSQOL) (Hines et al., 2015). Due to the convenience in definition and calculation, objective speech quality metrics are widely reported by an overwhelming majority of the literature.

Subjective Metrics. However, as pointed out by Reddy et al. (2019), the objective metrics may deviate from the experimental results in the subjective tests conducted by human beings. Therefore, it is necessary to introduce subjective metrics to better reflect the speech quality of human subjectivity. For example, the Deep Noise Suppression Challenge in 2022 (Dubey et al., 2022) is ranked according to the crowd-sourcing measure of the ITU-T Recommendation P.835 with Validation (ITU-T P.835) (Naderi & Cutler, 2020), comprised of three scores for each audio clip: speech quality (SIG), background noise quality (BAK), and overall quality (OVRL). It is likely, however, that *privacy-sensitive* content needs to be evaluated by human subjects in order to fully understand the extent of privacy needs.

Dynamic Blurring or Hiding of Shared Screen for Privacy

Overview of Dynamic Blurring or Hiding Shared Screen

Screen sharing is a typical operation during remote learning. Screen sharing includes the sharing of websites, documents, and videos. Sharing this content is useful to increase the effectiveness of remote learning. One challenge, however, is that the shared screen can include sensitive information (Lieberman, 2020). It is especially challenging to share only the content useful to the remote learning sessions. Because of this, it is common that during remote learning sessions, students sometimes are not willing to share their screens, which compromises the effectiveness of learning. In addition, on the teacher's side, not being able to selectively share a screen is also inefficient for learning because it takes time for a teacher to identify the content that they want to share. Meanwhile, a teacher's personal information may be exposed to all students during the trial-and-error process in finding the right content. As such, selective sharing of the needed content is quite crucial for successful remote learning in terms of both learning effectiveness and privacy conservation.

An effective strategy for a selective sharing of the needed content can be achieved through dynamic blurring or hiding of the shared screen. This process would blur or hide the privacy-critical information while keeping the needed content for others to view. Unfortunately, this concept is rather new, and there are no available applications able to achieve this goal yet. The following sections will outline the steps and fundamental techniques essential to implement dynamic blur and hiding of content for remote learning privacy.

Methods of Dynamic Blurring or Hiding a Shared Screen

The first necessary step to achieve dynamic blurring or hiding the shared screen is to recognize the text of the shared screen. "The text" refers to text anywhere on the computer screen. Through recognition of the text, it is feasible to blur or hide the privacy-sensitive content. For recognition of the screen text, optical character recognition (OCR) techniques can be used. OCR relies on machine learning to automatically recognize the natural language of a scene. Tesseract, originally developed by HP and became open source in 2006, is a well-established OCR app for this purpose (Smith, 2013). Tesseract relies on a series of steps in recognition of text. First, an image is put through an adaptive thresholding to remove non-interested features such as the background. It is followed by page layout analysis and word recognition passes. Typically, two passes are needed for recognition of the text. Post processing is immediately followed to correct the recognized text height and fuzzy text correction, as well as word bigram correction. Word bigram correction is a measure of word sequence in a sentence (Srinidhi, 2019). For example, in the sentence, "I need your help," "help" is the word most likely to follow the word "your." If the OCR gave some word other than "help," Tesseract would automatically correct it. As a commercial ready application, Tesseract is most convenient to be integrated to the remote learning environment to automatically recognize the shared screen text. Once the privacy-sensitive content is recognized, blurring or hiding operation of the shared screen can be executed for preserving privacy.

Recognition of text in the regular scene is not a challenging task for Tesseract. However, recognition of text under a natural scene such as the text on a wall or text on a coffee cup in the shared screen is rather challenging for Tesseract. It is because of this that text does not have regular shapes, sizes, and orientations. It is therefore difficult for Tesseract to recognize it. The situation becomes worse if the lighting conditions vary, such as in a dark environment. To tackle this condition, a more powerful OCR application needs to be created. Recent deep learning-based OCR research has started to tackle this problem. Among them, TextOCR has achieved success for this goal (Singh et al., 2021). TextOCR is a large-scale arbitrary shape text recognition application that has created a 900K large scale database with various sizes, shapes, and orientation texts. It uses faster region-based convolution neural network (RCNN) to localize the text (Girshick, 2015). A text extractor uses a segmentation proposal network to extract the text. The text extractor can extract arbitrary shape and orientation, therefore making it suitable for natural scene OCR text recognition (Liao et al., 2020). It then obtains the OCR text and embeds the text into vector (Hu et al., 2020). Consequently, a pointer-based network is used to organize the sequence of the words to ensure the semantic meaning of the recognized text (Singh et al., 2021). With a large database (900K) for training, also benefiting from the rigorous data processing pipeline, TextOCR is superior for recognition of text in natural scenes. It can achieve the goal of recognizing the shared screen text for preserving privacy in remote learning.

Recognition of the meaning of an image is also important for preserving privacy. For example, a student who has visited a gaming website may not want other students or the teacher to know the games that they have played. It is therefore expected that the gaming image needs to be recognized. For this purpose, classification of an image into *privacy-sensitive* and *non-privacy sensitive* is important. Classification of an image requires the collection and labeling of large-scale images. Image classification is a relatively simple task and well-studied. Therefore, the use of image classification for verification of image sensitivity is practical. The recent deep-learning revolution has largely improved image classification accuracy. State-of-the-art image classifiers have achieved over 85-90% accuracy and (Lu & Weng, 2007) . For CNN-based image classification, EfficientNet is considered superior in its performance and accuracy. EfficientNet uses a neural network search method to search for an ideal network structure. It balances the network depth, width, and feature space resolution. The results of the optimal search have made EfficientNet yield optimal classification. It has achieved 84% top-1 accuracy on the large-scale image database, ImageNet. Meanwhile, the model is seven times smaller and five times faster than Resnet-152 (Tan & Le, 2019). The great improvement of efficiency and small size of the model mean that EfficientNet offers practical real-world use for classification.

CNN-based image classification has been facing significant bottlenecks for further improvement of accuracy. Recent research on deep learning transformer image processing has further revolutionized image classification (Chen et al., 2021; Dosovitskiy et al., 2020; Zhai et al., 2022; Zhuang et al., 2021). In contrast to CNN-based classification, transformer-based image classification treats images as a series of patches. These patches are further tokenized with position encoding, followed by a transformer encoder to perform normalization and multiple head attention on these patches. Through the process, it is crucial that the multiple head attention extracts global features precisely to ensure the network attends to correct features of the image for image classification. It is proven that such global attention mechanisms are critical for classification accuracy (Zhai et al., 2022; Zhuang et al., 2021). Vision Transformer has shown its excellent performance in model scaling. Trained on the large-scale image database ImageNet, Vision Transformer shows excellent top-1 image classification accuracy of over 90%. Similar performance is also attained by training on a very large image database, where 3 billion images are used for image classification. Similarly, Vision Transformer shows a better performance than convolutional neural network-based classification (Zhai et al., 2022). Such improvement of classification accuracy demonstrates that classification of images for remote learning privacy preservation has become feasible.

It is also feasible that students may not want to show a private video to others. For example, they may wish to hide or blur a funny video about sports or a joke video about a lifestyle. Similarly, some students may share offensive videos which need to be disabled by the meeting host. For this purpose, video classification has become important. Video classification takes the video as input and classifies it into different categories. For privacy preservation in remote learning, the video needs to be classified into either privacy-sensitive or not-privacy-sensitive. Deep learning has achieved great progress for classifying videos. Among this work, convolutional neural network has been shown capable of classifying video with decent success (Karpathy et al., 2014). The convolutional model takes the frames of the videos and inputs them to a classification network, showing that the classification network can classify the video. It further shows that the video can be classified by using only one frame of the video. Pre-training the model on a large and more general video database is also shown to be helpful to improve the classification accuracy. Overall, the model is able to achieve 65% three-fold accuracy of classification, which is reasonable for practical video privacy classification (Karpathy et al., 2014).

To better classify video for privacy preservation, improved video classification accuracy is desirable. The root cause of the lack of video classification accuracy is the use of a single frame of video for video classification. Consequently, use of multiple frames for video classification is likely able to improve the classification accuracy. Research has shown success in capturing such temporal relationships of the video. It is known that the use of Stand-Alone Inter-Frame Attention can capture the intrinsic relation between frames and meanwhile attend to the correct features of the video (Long et al., 2022). Most importantly, the Inter-Frame Attention mechanisms can track video objects across video frames such that it is possible to more accurately classify videos. Results show that such inter-frame video attention can increase video classification accuracy from 65% to 75% for top-1 accuracy. Such improvement is meaningful for privacy preservation in remote learning to better classify videos.

Following the first step of recognizing text, image, and video and correctly classifying its privacy, the second step is to either blur or hide the privacy-sensitive content. For the purpose of blurring the shared content, an image filter is desired. The application of filters to blur privacy-sensitive content is relatively straightforward. Mostly, a filter is applied on the desired content to blur the specified region, achieving the goal. However, hiding *privacy-sensitive* content involves some work to crop the specified region. If the region is at the top or bottom, the crop will be easier to implement, but if the content is at the center of screen, cropping the privacy-sensitive content will yield a blank part representing missing content. Therefore, it will impact the quality of the shared screen. Under this condition, replacement of the privacy-sensitive content with other content is desired. The other content can be an icon or an image the students select so that it will be pleasant to view while also preserving privacy in remote learning.

Challenges of Dynamic Blurring or Hiding Shared Screen
It is worth noting that the blurring of specified *privacy-sensitive* regions may lead to discomfort during the remote learning session. Other students may regard the blurred content as a strange phenomenon. To alleviate this, hiding *privacy-sensitive* content may be more desirable. Meanwhile hiding the specified content is also problematic. Hiding content may introduce flickering of the screen if not implemented correctly, thereby also introducing visual discomfort. As such, care is needed to implement these techniques to ensure that remote learning can be conducted smoothly without any strange feelings associated with it.

Conclusion
This chapter has discussed three methods to preserve privacy in remote learning. The first method involves the use of a virtual background to replace the actual background. The virtual background technique has been widely adopted as the industry gold standard for video sessions in remote learning. While it has been widely used, most software including Zoom and Google Meet are not yet mature in terms of technology. Glitches still exist while using the virtual background, particularly when the actual background is dynamic rather than static, making it challenging to use this technique in remote learning. As such, a review of more state-of-the-art virtual background techniques is meaningful. It is expected that with these new techniques, it will become more feasible to conduct remote learning with a virtual background on/in an outdoor environment or highly dynamic indoor environment.

The second method proposes active noise suppression techniques for removal of background noises to preserve privacy. This chapter has systematically reviewed the conventional methods such as waveform spectrum and Gaussian distribution based statistical methods. Deep learning methods include spectrogram based convolutional neural network methods and Fourier domain based short-time Fourier transform methods. These state-of-the-art methods are proven rather effective in the suppression of background noises. The performance metrics including objective and subjective metrics are also given to evaluate the results of suppression. The advancement of audio and text analysis has shown it is practical to develop these background suppression models to effectively suppress *privacy-sensitive* sound to ensure the remote learning environment is free of privacy concerns.

The third method involves blurring or hiding sensitive shared content which is another critical task for preserving privacy. This chapter has reviewed methods for recognizing text, images, and videos. Through recognition of these types of content, it is feasible to either blur or hide the *privacy-sensitive* content. With state-of-the-art research, understanding the full text of the shared screen is feasible, therefore it is possible to recognize and understand the image and video to preserve privacy. Of course, there is other content such as animation GIF files and PDF attachments that are also *privacy-sensitive*. This content can be converted to either text, images, or videos first. Subsequently, content recognition can be performed to preserve privacy. The major concern of blurring or hiding content is the discomfort in viewing the blurred shared content, which can trigger suspicion from other viewers. As such, alternative operations such as hiding the content with another image may be preferable.

References

Bakkay, M. C., Rashwan, H. A., Salmane, H., Khoudour, L., Puig, D., & Ruicheck, Y. (2018). BSCGAN: Deep background subtraction with conditional generative adversarial networks. In *2018 25th IEEE International Conference on Image Processing (ICIP)* (pp. 4018–4022). IEEE. https://doi.org/10.1109/ICIP.2018.8451603

Beerends, J. G., Schmidmer, C., Berger, J., Obermann, M., Ullmann, R., Pomy, J., & Keyhl, M. (2013). Perceptual objective listening quality assessment (POLQA), the third generation ITU-T standard for end-to-end speech quality measurement part I—temporal alignment. *Journal of the Audio Engineering Society, 61*(6), 366–384.

Black, P. E. (2019 February 11). Manhattan distance. In *Dictionary of Algorithms and Data Structures* [online]. https://www.nist.gov/dads/HTML/manhattanDistance.html

Bu, Y., Zou, S., Liang, Y., & Veeravalli, V. (2018). Estimation of KL divergence: Optimal minimax rate. In *IEEE Transactions on Information Theory, 64*(4), 2648–2674. https://doi.org/10.1109/TIT.2018.2805844

Chen, X., Hsieh, C.-J., & Gong, B. (2021). When vision transformers outperform ResNets without pre-training or strong data augmentations. ArXiv Preprint. ArXiv:2106.01548v3. https://doi.org/10.48550/arXiv.2106.01548

Chuang, Y. Y., Curless, B., Salesin, D. H., & Szeliski, R. (2001). A Bayesian approach to digital matting. In *Proceedings of the 2001 IEEE Computer Society Conference on Computer Vision and Pattern Recognition.* IEEE. https://doi.org/10.1109/CVPR.2001.990970

Cioppa, A., Van Droogenbroeck, M., & Braham, M. (2020). Real-time semantic background subtraction. In *2020 IEEE International Conference on Image Processing (ICIP)*, (pp. 3214-3218). IEEE. https://doi.org/10.1109/ICIP40778.2020.9190838

Dosovitskiy, A., Beyer, L., Kolesnikov, A., Weissenborn, D., Zhai, X., Unterthiner, T., Dehghani, M., Minderer, M., Heigold, G., Gelly, S., Uszkoreit, J., & Houlsby, N. (2020 September 28). An image is worth 16×16 words: Transformers for image recognition at scale. In *ICRL 2021*, 1–21.

Dubey, H., Gopal, V., Cutler, R., Aazami, A., Matusevych, S., Braun, S., Eskimez, S. E., Thakker, M., Yoshioka, T., Gamper, H., & Aichner, R. (2022). Icassp 2022 deep noise suppression challenge. In *ICASSP 2022 – 2022 IEEE International Conference of Acoustics, Speech and Signal Processing (ICASSP)*, 9271–9275.

Ephraim, Y., & Malah, D. (1985). Speech enhancement using a minimum mean-square error log- spectral amplitude estimator. *IEEE Transactions on Acoustics, Speech, and Signal Processing, 33*(2), 443–445. https://doi.org/10.1109/TASSP.1985.1164550

Ephrat, A., Mosseri, I., Lang, O., Dekel, T., Wilson, K., Hassidim, A., Freeman, W. T., & Rubenstein, M. (2018). Looking to listen at the cocktail party: A speaker-independent audio-visual model for speech separation. *ACM Transactions on Graphics, 37*(4), 1–11. https://doi.org/10.1145/3197517.3201357

Feng, X., Liang, X., & Zhang, Z. (2016, September 17). A cluster sampling method for image matting via sparse coding. In: B. Leibe, J. Matas, N. Sebe, & M. Welling (eds) *Computer Vision—ECCV 2016.* https://doi.org/10.1007/978-3-319-46475-6_13

Gastal, E. S., & Oliveira, M. M. (2010, June 07). Shared sampling for real-time alpha matting. *Computer Graphics Forum, 29*(2), 575–584. https://doi.org/10.1111/j.1467-8659.2009.01627.x

Girshick, R. (2015). Fast R-CNN. In *2015 IEEE International Conference on Computer Vision (ICCV)*, 1440–1448. http://doi.org/10.1109/ICCV.2015.169.

Hani, M. (2020). Best practices for implementing remote learning during a pandemic. *The Clearing House: A Journal of Educational Strategies, Issues and Ideas, 93*(3), 135–141. https://doi.org/10.1080/00098655.2020.1751480

Hines, A., Skoglund, J., Kokaram, A. C., & Harte, N. (2015 May 17). ViSQOL: An objective speech quality model. *EURASIP Journal on Audio, Speech, and Music Processing, 13*. https://doi.org/10.1186/s13636-015-0054-9

Hu, R., Singh, A., Darrell, T., & Rohrbach, M. (2020). Iterative answer prediction with pointer-augmented multimodal transformers for TextVQA. In *2020 IEEE/CVF Conference on Computer Vision and Pattern Recognition (CVPR)*, 9989–9999. https://doi.org/10.1109/CVPR42600.2020.01001

Johnson, D. H. (2006). Signal to noise ratio. In *Scholarpedia, 1*(12). http://dx.doi.org/10.4249/scholarpedia.2088

Karpathy, A., Toderici, G., Shetty, S., Leung, T., Sukthankar, R., & Fei-Fei, L. (2014). Large-scale video classification with convolutional neural networks. In *2014 IEEE Conference on Computer Vision and Pattern Recognition*, 1725–1732. https://doi.org/10.1109/CVPR.2014.223

Koyama, Y., Vuong, T., Uhlich, S., & Raj, B. (2020). Exploring the best loss function for DNN-based low-latency speech enhancement with temporal convolutional networks. ArXIV Preprint. ArXiv:2005.11611v3. https://doi.org/10.48550/arXiv.2005.11611

Levin, A., Lischinski, D., & Weiss, Y. (2007, December 18). A closed-form solution to natural image matting. *IEEE Transactions on Pattern Analysis and Machine Intelligence, 30*(2). https://doi.org/10.1109/TPAMI.2007.1177

Li, C., & Lalani, F. (2020, April 29). *The COVID-19 pandemic has changed education forever. This is how.* World Economic Forum. https://www.weforum.org/agenda/2020/04/coronavirus-education-global-covid19-online-digital-learning/

Liao, M., Pang, G., Huang, J., Hassner, T., & Bai, X. (2020 August). Mask TextSpotter v3: Segmentation proposal network for robust scene text spotting. In *Computer Vision – ECCV 2020: 16th European Conference, Glasgow, UK, August 23-28, 2020, Proceedings, Part XI*, 706–722. https://doi.org/10.1007/978-3-030-58621-8_41

Lieberman, J. (2020 May 27). Following pornographic bookmark incident, instructor says UM pushed him to resign. *The Miami Hurricane*. https://www.themiamihurricane.com/2020/05/27/following-pornographic-bookmark-incident-instructor-says-um-pushed-him-to-resign/

Long, F., Qiu, Z., Pan, Y., Yao, T., Luo, J., & Mei, T. (2022). Stand-alone inter-frame attention in video models. In *2022 IEEE/CVF Conference on Computer Vision and Pattern Recognition (CVPR)*, 3182–3191. https://doi.org/10.1109/CVPR52688.2022.00319

Lotter, T., & Vary, P. (2005). Speech enhancement by map spectral amplitude estimation using a super-gaussian speech model. *EURASIP Journal on Advances in Signal Processing*. 354850. https://doi.org/10.1155/ASP.2005.1110

Lu, D., & Weng, Q. (2007). A survey of image classification methods and techniques for improving classification performance. *International jJournal of Remote sSensing, 28*(5), 823-870. https://doi.org/10.1080/01431160600746456

Luo, Y., & Mesgarani, N. (2019). Conv-TasNet: Surpassing ideal time–frequency magnitude masking for speech separation. *IEEE/ACM Transactions on Audio, Speech, and Language Processing, 27*(8), 1256–1266. https://doi.org/10.1109/TASLP.2019.2915167

Naderi, B., & Cutler, R. (2020). A crowdsourcing extension of the ITU-T recommendation p.835 with validation. ArXiv Prepublication. ArXiv:2010.13200v1. https://github.com/microsoft/P.808

Nocerino, N., Soong, F., Rabiner, L., & Klatt, D. (1985). Comparative study of several distortion measures for speech recognition. In *ICASSP '85. IEEE International Conference on Acoustics, Speech, and Signal Processing*, 25–28. https://doi.org/10.1109/ICASSP.1985.1168478

Pandey, A., & Wang, D. (2019). TCNN: Temporal convolutional neural network for real-time speech enhancement in the time domain. In *ICASSP 2019 – 2019 IEEE International Conference on Acoustics, Speech and Signal Processing (ICASSP)*, 6875–6879. https://doi.org/10.1109/ICASSP.2019.8683634

Patil, P. W., Dudhane, A., & Murala, S. (2021). Multi-frame recurrent adversarial network for moving object segmentation. In *2021 IEEE Winter Conference on Applications of Computer Vision (WACV)* (pp. 2301-2310). IEEE. https://doi.org/10.1109/WACV48630.2021.00235

Reddy, C. K. A., Shankar, N., Bhat, G. S., Charan, R., & Panahi, I. (2017). An individualized super-Gaussian single microphone speech enhancement for hearing aid users with smartphone as an assistive device. *IEEE Signal*

Processing Letters, 24(11), 1601–1605.
https://doi.org/10.1109/LSP.2017.2750979

Reddy, C. K., Beyrami, E., Pool, J., Cutler, R., Srinivasan, S., & Gehrke, J. (2019). A scalable noisy speech dataset and online subjective test framework. In *Proc. Interspeech 2019*, 1816–1820. http://dx.doi.org/10.21437/Interspeech.2019-3087

Rix, A., Beerends, J., Hollier, M. P., & Hekstra, A. P. (2001). Perceptual evaluation of speech quality (PESQ)-a new method for speech quality assessment of telephone networks and codecs. In *2001 IEEE International Conference on Acoustics, Speech, and Signal Processing, 2*, 749–752. https://doi.org/10.1109/ICASSP.2001.941023

Sengupta, S., Jayaram, V., Curless, B., Seitz, S., & Kemelmacher-Shlizerman, I. (2020). Background matting: The world is your green screen. In *2020 IEEE/CVF Conference on Computer Vision and Pattern Recognition (CVPR)* (pp. 2288-2297). IEEE. https://doi.org/10.1109/CVPR42600.2020.00236

Singh, A., Pang, G., Toh, M., Huang, J., Galuba, W., & Hassner, T. (2021). TextOCR: Towards large-scale end-to-end reasoning for arbitrary-shaped scene text. In *2021 IEEE/CVF Conference on Computer Vision and Pattern Recognition (CVPR)*, 8798–8808. https://doi.org/10.1109/CVPR46437.2021.00869

Smith, R. W. (2013 February 04). History of the Tesseract OCR engine: What worked and what didn't. In *Proc. SPIE 8658, Document Recognition and Retrieval XX, 865802*. https://doi.org/10.1117/12.2010051

Srinidhi, S. (2019 November 27). *Understanding word n-grams and n-gram probability in natural language processing*. Towards Data Science. https://towardsdatascience.com/understanding-word-n-grams-and-n-gram-probability-in-natural-language-processing-9d9eef0fa058

Srinivasan, S., Samuelsson, J., & Kleijn, W. B. (2007). Codebook-based Bayesian speech enhancement for nonstationary environments. *IEEE Transactions on Audio, Speech, and Language Processing, 15*(2), 441–452. https://doi.org/10.1109/TASL.2006.881696

Sultana, M., Mahmood, A., & Jung, S. K. (2022). Unsupervised moving object segmentation using background subtraction and optimal adversarial noise sample search. *Pattern Recognition, 129*, 108719. https://doi.org/10.1016/j.patcog.2022.108719

Sun, J., Jia, J., Tang, C.-K., & Shum, H.-Y. (2004, August 01). Poisson matting. In *SIGGRAPH '04: ACM SIGGRAPH 2004 Papers* (pp. 315-321). https://doi.org/10.1145/1186562.1015721

Tan, M., & Le, Q. (2019). EfficientNet: Rethinking model scaling for convolutional neural networks. In *Proceedings of the 36th International*

Conference on Machine Learning, PMLR 97:6105–6114. https://proceedings.mlr.press/v97/tan19a.html

Tezcan, M. O., Ishwar, P., & Konrad, J. (2021). BSUV-net 2.0: Spatio-temporal data augmentations for video-agnostic supervised background subtraction. *IEEE Access, 9*, 53849–53860. https://doi.org/10.1109/ACCESS.2021.3071163

Wang, Q., Li, S., Wang, C., Dai, M. (2020). Effective background removal method based on generative adversary networks. *Journal of Electronic Imaging, 29*. http://doi.org/10.1117/1.JEI.29.5.053014

Watcharasupat, K. N., Nguyen, T. N. T., Woon-Seng, G., Shengkui, Z., & Ma, B. (2022). End-to-end complex-valued multidilated convolutional neural network for joint acoustic echo cancellation and noise suppression. In *ICASSP 2022 – 2022 IEEE International Conference on Acoustics, Speech and Signal Processing (ICASSP)*, 656–660. https://doi.org/10.1109/ICASSP43922.2022.9747034

Westhausen, N. L., & Meyer, B. T. (2020). Dual-signal transformation LSTM network for real-time noise suppression. In *Proc. Interspeech 2020*, 2477–2481. http://dx.doi.org/10.21437/Interspeech.2020-2631

Wolfe, P. J., & Godsill, S. J. (2001). Simple alternatives to the Ephraim and Malah suppression rule for speech enhancement. In *Proceedings of the 11th IEEE Signal Processing Workshop on Statistical Signal Processing (Cat. No. 01TH8563)* (pp. 496–499). https://doi.org/10.1109/SSP.2001.955331

Xia, Y., Bruan, S., Reddy, C. K. A., Dubey, H., Cutler, R., & Tashev, I. (2020). Weighted speech distortion losses for neural-network-based real-time speech enhancement. In *ICASSP 2020 – 2020 IEEE International Conference on Acoustics, Speech, and Language Processing (ICASSP)*, 871–875. https://doi.org/10.1109/ICASSP40776.2020.9054254

Xu, N., Price, B., Cohen, S., & Huang, T. (2017, November 09). Deep image matting. In *2017 IEEE Conference on Computer Vision and Pattern Recognition (CVPR). https://doi.org/10.1109/CVPR.2017.41*

Xu, Y., Du, J., Dai, L.-R., & Lee, C.-H. (2014). A regression approach to speech enhancement based on deep neural networks. *IEEE/ACM Transactions on Audio, Speech, and Language Processing, 23*(1), 7–19. https://doi.org/10.1109/TASLP.2014.2364452

Yang, Y., Cordeil, M., Beyer, J., Dwyer, T., Marriott, K., & Phister, H. (2020, October 13). Embodied navigation in immersive abstract data visualization: Is overview+detail or zooming better for 3D scatterplots? *IEEE Transactions on Visualization and Computer Graphics, 27*(2), 1214–1224. IEEE. https://doi.org/10.1109/TVCG.2020.3030427

Yin, D., Luo, C., Xiong, Z., & Zeng, W. (2020, April 03). PHASEN: A phase-and-harmonics-aware speech enhancement network. In *Proceedings of the AAAI Conference on Artificial Intelligence, 34*, 9458–9465. https://doi.org/10.1609/aaai.v34i05.6489

Zhai, X., Kolesnikov, A., Houlsby, N., & Beyer, L. (2022). Scaling vision transformers. In *2022 IEEE/CVF Conference on Computer Vision and Pattern Recognition (CVPR)*, 1204–1213. https://doi.org/10.1109/CVPR52688.2022.01179

Zhai, X., Wang, X., Mustafa, B., Steiner, A., Keysers, D., Kolesnikov, A., & Beyer, L. (2022). LiT: Zero-shot transfer with locked-image text tuning. In *2022 IEEE/CVF Conference on Computer Vision and Pattern Recognition (CVPR)*, 18102–18112. https://doi.org/10.1109/CVPR52688.2022.01759

Zhuang, J., Gong, B., Yuan, L., Cui, Y., Adam, H., Dvornek, N. C., Tatikonda, S., Duncan, J. S., & Liu, T. (2021). Surrogate gap minimization improves sharpness-aware training. In *ICLR 2022 Conference*, 1–24.

About the Authors

Editors

Denise FitzGerald Quintel (she/her) is an Associate Professor and Discovery Services Librarian at Middle Tennessee State University. Her research involves information-seeking behavior, user-centered design, web analytics, and privacy. She has published work in the *Journal of Academic Librarianship, Issues in Science and Technology Librarianship,* and *Information Technology and Libraries.* Denise is a member of the Asian Pacific American Library Association (APALA) and the Library Freedom Project. She currently serves as an Assistant Editor for *Weave: Journal of Library User Experience.*

Amy York is an Associate Professor and User Services Librarian at the James E. Walker Library at Middle Tennessee State University. Her work currently focuses on using instructional technology to teach information literacy and research skills, but she has also worked with and written about distance learning services, web services, and digital collections. Amy has published articles in *OCLC Systems and Services, College & Research Library News, Collaborative Librarianship,* and *the Journal of Library Administration.* She is also a past editor and current reviewer for *Tennessee Libraries,* the journal of the Tennessee Library Association.

Contributors

Emma Antobam-Ntekudzi is a Reference & Instruction Librarian at Bronx Community College. In her role, Emma helps students, staff, and teaching faculty with research needs and provides library instruction sessions. Her interest in librarianship began during her time as a Library Associate at the New York Botanical Garden's LuEsther T. Mertz Library. She received an MLS and an MA in Urban Affairs from Queens College. In 2018, she was honored as the NYLA-NYBLC Diversity & Inclusion Scholar, and in 2021, she was awarded the SLA James M. Matarazzo Rising Star Award. Her personal interests include reading, journaling, and spending time with family.

Christian Barborini is an MS student in the School of Population and Public Health at the University of British Columbia, as well as a researcher at the BC Centre on Substance Use (BCCSU) and the Centre for Gender & Sexual Health Equity (CGSHE). Christian's research, co-supervised by Dr. Rod Knight and

Dr. Mark Gilbert, focuses on cannabis use in relation to the gender experiences of trans and non-binary youth. Prior to this, they completed their Honours BS at McMaster University in Biochemistry and Biomedical Sciences. In addition to their science background, Christian's research is informed by their years of experience working and organizing within 2S/TLGBQIA+ community. They are committed to approaching their research from an intersectional, community-based lens that accounts for various forms of marginalization experienced by 2S/TLGBQIA+ youth who use substances.

Maddie Brockbank (she/her) is a PhD Candidate and Vanier Scholar in the School of Social Work at McMaster University. Maddie has her Bachelor of Social Work (2019) and Master of Social Work (2020) from McMaster. Her research, practice experience, and community organizing initiatives have been in the area of anti-violence work with men, specifically in exploring the links between sexual violence prevention, anti-carceral feminisms, and engaging men. Additionally, she has research experience in the areas of houselessness, disability, social systems, curriculum development, and creating safety for marginalized students in university pedagogy. Maddie has been recognized for her academic excellence and community leadership as a recipient of the Young Woman of Distinction Award (YWCA Hamilton), the President's Award for Excellence in Student Leadership (McMaster University), a Hamilton Hero award (Hamilton Ticats), a Women Who Rock award (EMPOWER Strategy Group), and a Mary Keyes Award for Outstanding Service and Leadership (McMaster University).

Samantha Clarke is the Grants Coordinator for the Office of the Vice-Provost, Teaching and Learning at McMaster University. She developed and implemented an institutional granting opportunity for instructors who seek to improve teaching and learning, with a focus on initiatives to support student engagement, belonging, and wellness. She also supports institutional applications to several external granting opportunities. Sam recently completed her PhD in medical history, examining how the Cold War atmosphere of distrust shaped the reception of novel vaccines to prevent polio in divided Germany. In her spare time, she is an avid gardener, embroiderer, and camper.

Christina M. Cobb is an Assistant Professor in the University Studies Department teaching mathematics at Middle Tennessee State University. Dr. Cobb works to give all her students the tools needed to be successful in her mathematics courses and beyond. She teaches with so much passion and energy that her students love to come to class. Dr. Cobb was awarded one of the Outstanding Teaching awards at MTSU in fall 2019. She has presented numerous times nationally about innovative ways to teach mathematics and serves as the co-chair of the Mathematics Network for the National Organization for Student Success (NOSS). Dr. Cobb was a co-principal investigator for the 2019 TBR SERS grant "Strategies to Enhance African

American Males in a Prescribed Mathematics Course Success Rates." Dr. Cobb has a great rapport with her students and was recently named the Inclusive Teaching Fellow with the LT&ITC at MTSU.

Angela Dixon currently works as a Reference & Instruction Librarian at George State University Library – Dunwoody Campus. She holds an A.A.S. in Electronics Engineering and a B.S. in Technical & Professional Communication. Angela obtained her MLIS from Valdosta State University. Her research interests include digital privacy issues that directly impact and affect BIPOC communities. Her passion is working with first- and second-generation college students. She has presented at local, state, and national conferences. Angela holds active membership in Georgia Library Association (GLA), American Library Association (ALA), and Library Freedom Project (LFP).

Wil Prakash Fujarczuk is an enthusiastic educator guided by critical pedagogy, intersectional feminism, and anti-oppression, and he is committed to facilitating transformative learning opportunities for participants. He is the manager of the sexual violence prevention education program at McMaster University's Equity and Inclusion Office and the resource person for the gender and sexuality working group of the President's Advisory Committee on Building an Inclusive Community. Wil also has a consent-educating drag persona named Unita Ask. He is the co-chair of the Canadian Association of College & University Student Services' Sexual Violence Prevention and Response Community of Practice, a member of the City of Hamilton's LGBTQ Advisory Committee, a qualified safeTALK trainer with LivingWorks, and an executive member of the Canadian Association for the Prevention of Discrimination and Harassment in Higher Education. Wil holds a BS in Biology (McMaster University), a BEd (University of Western Ontario), and an MA in Peace Education (UN-mandated University for Peace).

Renata Hall lives all things social justice! With an educational background in psychology and biology from Dalhousie University and social work from McMaster University, Renata looks to shake the room through her love of building community connections, challenging the status quo, and amplifying the voices of the margins. Seasoned in racialized peer support, counseling, and teaching through McMaster University, housing and homelessness support with the YWCA Hamilton, as well as food equity and community network building through her grassroots initiative StreetEatzHamOnt, Renata brings passion and a sharp Black Feminist and Critical Race lens to every conversation and every table. As she is comfortably situated in the education and policy sector, Renata looks to bring a liberating, empowering, and critically conscious force to the Hamilton Community at large.

James Hamby is the Associate Director of the Margaret H. Ordoubadian University Writing Center at Middle Tennessee State University, where he also teaches courses in composition and literature. He serves on the editorial boards of *Praxis*, *Southern Discourse In the Center*, and *WLN: A Journal of Writing Center Scholarship*. He is currently serving his second term as Tennessee Representative on the executive board of the Southeastern Writing Center Association and has presented writing center research at many conferences.

Meredith Anne (MA) Higgs is an Associate Professor of University Studies at Middle Tennessee State University teaching undergraduate prescribed mathematics and undergraduate and graduate Professional Studies coursework. Her research and educational agenda emphasize retention and success of students who are non-traditional, academically at-risk, diverse, and enrolled in distance education. She emphasizes High Impact Practices (HIPs), active learning, and gamification in her classroom and student engagement activities. She is the Co-chair of the National Organization for Student Success Mathematics Network, has co-authored OER works in addition to journal articles and teaching contributions, and is a well-regarded, high-energy national conference presenter. For her work, she has been recognized with the MTSU Outstanding Professor award, the national Gladys R. Shaw Award for Outstanding Service and Support for Student Success, and the Association for Continuing Higher Education South: Outstanding Faculty Award.

Albert Kagan has contributed to the application and practice of using information-based technology for business organizations from a security standpoint. He has served as a subject matter expert for Homeland Security issues for over fifteen years. Professor Kagan has been an investigator on over 60 external research grants that have attained approximately $20 million in external funding. He has participated in projects that have total grant funding of over $38 million. He has held editorial or reviewer positions at numerous academic journals and has received two awards from professional associations for his research activities. In 2015, Professor Kagan was a finalist for the Jefferson Fellows Program sponsored by the U S Department of State. He has published nearly 190 referred articles and presentations.

Joseph Kennedy is a National Board for Professional Teaching Standards-certified mathematics educator with an M.Ed. in Instructional Design. As the Instructional Designer for Concordia College, he co-led the institution's pivot to distance learning during the COVID-19 pandemic and is the primary administrator of the college's Learning Management System. He also serves as an intermittent adjunct faculty member in the Education and Mathematics departments and was recognized by his peers with the Flatt Distinguished Service award. A former national champion forensicator, he has coached the college's nationally ranked speech team. Prior to entering higher education, he taught high school mathematics and computer programming, co-developed a new mathematics curriculum for his district, and pioneered competency-based courses for students who struggled in traditional classrooms. He has also served as a labor mediator between the Fargo Education Association and Fargo Public Schools and enjoys being a poll worker during elections.

Hannah Lee is a double alumna of UCLA, where she obtained her BA in English and MLIS. As the Discovery & Systems Librarian for California State University, Dominguez Hills, she collaborates with other library staff and faculty to develop policies, procedures and workflows that enhance access and discoverability of the library's collections, web service integrations, and support the curation of the library's digital content. Some of her interests include copyright, intellectual property, cybernetics, book arts, mentoring, and volunteering for non-profit organizations.

Kimberly Looby is the Instruction and Information Literacy Librarian at the University of North Carolina at Charlotte. She has a Master of Library and Information Science from the University of Illinois Urbana-Champaign, a Master of Arts in Historical Administration from Eastern Illinois University, and a Bachelor of Arts in Anthropology from the University of Illinois Urbana-Champaign. Her professional interests include invisible disability in librarianship, first year students, and accessible assessment and data visualization for all librarians.

Lei Miao is an Associate Professor in the Department of Engineering Technology at MTSU. He earned his PhD from Boston University and his master's and bachelor's degrees from Northeastern University of China, in 2006, 2001, and 1998, respectively. From 2006 to 2009, he was with Nortel Networks in Billerica, MA. From 2009 to 2011, he was a visiting assistant professor at the University of Cincinnati. From 2011 to 2014, he was a wireless networking and software engineer with NuVo Technologies/Legrand North America. From 2014 to 2015, he was an assistant professor at the State University of New York at Farmingdale. He has almost 20 years R&D experience in system control and optimization, information systems, and Intelligent Transportation Systems (ITS). He has near 40 publications in referred conferences and journals. He is a member of the technical program committee of many international conferences and the guest editor of a Special Issue "Advances in Smart City and Intelligent Transportation Systems" of the MDPI journal *Sustainability*.

Jonathan B. Moore is the User Experience Librarian for J. Murrey Atkins Library at University of North Carolina at Charlotte. He holds an MSLS from UNC Chapel Hill in Chapel Hill, North Carolina. His work seeks to empower library users in the design and assessment of library spaces, resources, and services. He researches the behavior of library users during information seeking, the design and use of informal learning environments, and the role of library services in supporting social justice. His previous work has been published in *Technical Services Quarterly* and presented at the Triad Academic Library Association and NC LIVE annual conferences.

Jennifer (Jenny) Reichart is a nationally recognized Holmes Scholar and DEI (diversity, equity, and inclusion) leader in higher education. Her research examines complex equity problems affecting underserved students using design and systems thinking. As a practitioner, she has served as a counselor and community leader creating open pathways to, from, and within higher education. She has worked in higher education as a supervisor, faculty member, senior-level administrator, and executive leader for the past fifteen years. She is an expert in trauma-informed/resilience-focused instruction, a trailblazer in faculty and student self-care practices, and a certified emotional intelligence life coach in the area of higher education executive leadership. In her current position as Executive Director of Diversity, Equity, Inclusion, and Belonging at Richland Community College, she serves as an internal organizational culture consultant, community thought leader on the President's Cabinet, and as the college's Title IX, ADA, EEOC, Grievance, and Compliance Officer.

Bridgette T. Sanders is the Social Sciences Research & Instructional Services Librarian at J. Murrey Atkins Library, University of North Carolina Charlotte in Charlotte, North Carolina. She received her MLS from Atlanta University in Atlanta, Georgia. Her research interests include diversity, equity, and inclusion, reference and instruction, virtual reference, and library spaces. She is the co-editor of the ALA editions book *Making the Most of Your Library Career*. She has presented at several conferences including the EDUCAUSE West/Southwest Regional Conference, the Library 2.012 Worldwide Virtual Conference, and the EdMedia: World Conference on Educational Media & Technology. Her recent presentations include the REFORMA Virtual Conference, LibLearnX Conference, AAC&U 2022 Conference on Diversity, Equity, and Student Success, and the 2022 NC LIVE Annual Conference.

Amy Stalker is an Associate Department Head at the Dunwoody campus of the Georgia State University Library. She holds a B.A. in Russian Area Studies from Wittenberg University and an MLIS from Valdosta State University. Amy is an active member of the American Library Association (ALA) and the Association of College & Research Libraries (ACRL). Her research interests include misinformation, digital privacy, and management approaches to employee development and morale. She has presented at local, state, and national library conferences, as well the Georgia Political Science Association Annual Conference and GSU's Conference on Scholarly Teaching: Global Conversations in Higher Education. She has co-authored several articles, book reviews, and book chapters.

Rebecca Taylor is an elementary education teacher in Chattanooga, Tennessee. She has taught three years in fifth grade but has worked with and observed all elementary grade-levels. She has a passion for education and is constantly working to improve the learning environment for all students. Taylor also has a passion for opportunities to help teachers facing burnout and helping them find opportunities to grow as a teacher and manage stress. She has a bachelor's degree in elementary education, a master's degree in educational theories and practices, and is working through an Assessment, Learning, and Student Success doctoral program at Middle Tennessee State University.

Kristen Vogt Veggeberg is a nonprofit director, scholar, and writer from the south side of Chicago. She holds a BA in Medieval Studies (Honors) from the University of Oregon, an MPA in Public Administration from Southern Illinois University, and a PhD in Curriculum and Instruction from the University of Illinois at Chicago. Her research is on informal STEAM education, and her dissertation focused on discourses of equity amongst museum educators. She has been published in *Anthropology and Education Quarterly*, *The National Teaching and Learning Forum*, and *Women in Higher Education*, amongst other publications. Vogt Veggeberg currently oversees a multi-state STEAM and career exploration program for a major nonprofit, as well as serving as a Science Communication Research Associate at the University of Oregon. You can visit her at https://linktr.ee/krisveevee.

Yimeng Wang is a J.D. candidate (Class of 2024) and a Root-Tilden-Kern Public Interest Scholar at NYU School of Law. Born in 西安, Yimeng grew up on Treaty 13 land, the Traditional Territories of the Anishinaabe peoples, Haudenosaunee Confederacy, and Mississaugas of the Credit First Nation. They graduated from the Arts and Science Program at McMaster University in 2021. During undergrad, they were involved with the Women and Gender Equity Network (WGEN), a student-run peer support service centering the needs of survivors of sexual violence, folks under the trans umbrella, and all individuals who experience gender-based oppression. They held the title of Coordinator in the 2020-2021 academic year. As a community member, they have organized with speqtrum (Hamilton), SACHA (Hamilton), Survived and Punished NY, and Red Canary Song (NY).

Sarah Whitwell is the Experiential Programming and Outreach Manager for the Faculty of Humanities at McMaster University. She completed her PhD in History at McMaster University and has several years of practical teaching experience in addition to her work in the field of teaching and learning. Sarah's teaching philosophy emphasizes the importance of active and inquiry-based learning, as well as cultivating an inclusive and accessible classroom environment where everyone feels as safe and comfortable as possible.

Aimin Yan is a Professor and PhD supervisor at Shanghai Normal University of China. She completed her PhD at Shanghai Institute of Optics and Fine Mechanics (SIOM) of the Chinese Academy of Sciences (CAS). Her research interests are information optics, optical image encryption, and 3D laser imaging. She has published more than 50 papers and applied for 15 invention patents. She teaches optoelectronics, Laser Principles & iTechnology, and Experiments of Modern Physics.

Hongbo Zhang is an Assistant Professor at Middle Tennessee State University. He completed his PhD at Virginia Tech. His research interests are Computer Vision, Robotics, and 3D Imaging. He has co-authored more than 70 publications. He has also mentored two PhD students and five master's students. He is a panelist for NASA Development and Advancement of Lunar Instrumentation program. He is the topic editor for *Chinese Optics Letters*. He is also the guest editor for *Electronics*. He teaches Computer Vision, Human Computer Interaction, Control, and Electronics.

Jia-Xing Zhong graduated from Sun Yat-sen University with a B. Eng. degree in Computer Science and Technology in 2017. He received an M. NatSci. Degree in Computer Applied Technology from Peking University in 2020, under the supervision of Prof. Ge Li. Currently, he is a PhD student (Oct. 2020 –) in the Department of Computer Science at the University of Oxford, supervised by Prof. Niki Trigoni and Prof. Andrew Markham. His main research interests are machine learning, computer vision, and robotics.

Acknowledgements

Generous support has been provided by the following individuals and organizations:

External Reviewers

For the entire book
Flower Darby
Bonnie Tijerina

For chapter(s)
Tamim Arif
Lanier Basenberg
Karen Reed
Robert Wilson

Middle Tennessee State University

Operations and financial support from Walker Library include:
Jayme Brunson, Administration
Anna Mills, Digital Scholarship Initiatives
Kathleen Schmand, Dean
Clay Trainum, Administration

Consultations and operational support from:
College of Business
Contract Office
Office of the University Counsel

Index

Blackboard, 15, 18, 21, 193
bullies *see* bullying
bullying, 69, 75, 76, 80-88
bystanders, 84-85, 88
cheating, 163, 169, 193
children, 80, 82-83, 87-88, 120, 135, 147, 182, 189
cyberattacks, 21
cyberbullying, 80-88
cybersecurity, 24, 155-156, 159
data breach(es), 19-21, 24
Discord, 47, 180, 184
doxing, 74
elementary school, 80-81, 87
essential employees, 143
essential personnel
    see essential employees
essential workers
    see essential employees
faculty training, 164
FERPA, 18-19, 22-23, 51, 115, 179-180
FOIA *see* Freedom of Information Act
Freedom of Information Act 73, 77
Google, 16, 19-20, 47, 50, 53-54, 56, 157, 159-160, 181, 200
guidance counselors, 85-88
hacking, 14, 17, 20, 74, 158-159
harassment, 31, 41, 71, 75-76, 80-81, 83, 87

learning management software, 15, 123-124, 126, 164-165, 167-172, 175
LMS *see* learning management software
malware, 19-20
Microsoft teams, 21, 34, 36, 47, 143, 148, 194
online identity, 72
personal identifiable information, 20-21, 74
ransomware, 21-22, 75
social media, 18-19, 24, 33, 50, 53-54, 69, 71-73, 75-76, 80-83, 86, 88, 93, 155, 184-185
spyware, 20, 125
suicide, 83-84
teens, 82, 86
test proctoring, 144, 147-148, 164
testing, 37, 103, 147, 174, 195
third-party vendors, 16-18, 22-23
TikTok, 82-83, 147
tutors, 179-186
UDL *see* Universal Design for Learning
Universal Design for Learning, 16, 129-130, 133-134, 136-137, 139, 164, 167-168
victims, 75, 83-88, 121
WCONLINE, 179-181
Zoom, 17-18, 21, 33-35, 38, 42, 47, 69, 118, 143-146, 148, 155, 170-172, 180-181, 194, 200
Zoom fatigue, 145-146

www.ingramcontent.com/pod-product-compliance
Lightning Source LLC
Chambersburg PA
CBHW070532090426
42735CB00013B/2952